In-Situ and Operando Probing of Energy Materials at Multiscale Down to Single Atomic Column—The Power of X-Rays, Neutrons and Electron Microscopy

MATERIALS RESEARCH SOCIETY
SYMPOSIUM PROCEEDINGS VOLUME 1262

In-Situ and Operando Probing of Energy Materials at Multiscale Down to Single Atomic Column—The Power of X-Rays, Neutrons and Electron Microscopy

Symposium held April 5–9, 2010, San Francisco, California

EDITORS:

Chongmin Wang
Pacific Northwest National Laboratory,
Richland, Washington, U.S.A.

Artur Braun
Swiss Federal Laboratories for Materials
Science and Technology
Dubendorf, Switzerland

Niels de Jonge
Vanderbilt University
Nashville, Tennessee, U.S.A.

Jinghua Guo
Berkeley National Laboratory
Berkeley, California, U.S.A.

Rafal E. Dunin-Borkowski
Technical University of Denmark
Lyngby, Denmark

Randall E. Winans
Argonne National Laboratory
Lemont, Illinois, U.S.A.

Helmut Schober
Institut Laue-Langevin
Grenoble, France

Materials Research Society
Warrendale, Pennsylvania

CAMBRIDGE UNIVERSITY PRESS
Cambridge, New York, Melbourne, Madrid, Cape Town,
Singapore, São Paulo, Delhi, Mexico City

Cambridge University Press
32 Avenue of the Americas, New York NY 10013-2473, USA

Published in the United States of America by Cambridge University Press, New York

www.cambridge.org
Information on this title: www.cambridge.org/9781107406681

Materials Research Society
506 Keystone Drive, Warrendale, PA 15086
http://www.mrs.org

© Materials Research Society 2010

First published 2010
First paperback edition 2012

Single article reprints from this publication are available through
University Microfilms Inc., 300 North Zeeb Road, Ann Arbor, MI 48106

CODEN: MRSPDH

ISBN 978-1-605-11239-8 Hardback
ISBN 978-1-107-40668-1 Paperback

CONTENTS

H2 STORAGE AND HYDROGEN IN SOLIDS I

H2 STORAGE AND HYDROGEN IN SOLIDS II

PEM FUEL CELLS AND ELECTROCATALYSIS

POSTER SESSION

BATTERIES AND PHOTOSYNTHESIS

SOLAR CELLS

*Invited Paper

POROUS MEDIA AND DISORDERED SYSTEMS

PREFACE

The functionalities of materials and devices have their origin typically in structural properties at the molecular scale and mesoscale, and may extend to the macroscale. Basic understanding of these interrelations requires detailed analysis of the structural properties of materials. There has been tremendous progress in the development of materials characterization methods.

This volume contains papers presented at Symposium V, "*In Situ* Transmission Electron Microscopy and Spectroscopy," and Symposium W, "Diagnostics and Characterization of Energy Materials with Synchrotron Neutron Radiation," both held April 5–9 at the 2010 MRS Spring Meeting in San Francisco, California. The purpose of these two symposia was to bring together a broad coalition of scientists from the fields that engaged in the use of electron microscopy and spectroscopy, neutrons, and synchrotron x-ray methods to probe into the structure and properties of materials. The symposia provided an ideal platform for discussing state-of-art techniques based on electrons, photons and neutrons for exploring the structure and property relationship of energy materials and nanomaterials.

One of the major highlights of these two symposia is the potential developments on the next generation *in-situ* and *ex-situ* techniques for materials science in general. There is a broad interest in understanding chemical transformations on surfaces, in materials interfaces, and in many areas of nanoscience, including catalysis, energy materials and environmental science. Advanced tools to obtain site-specific information in realistic reaction environments are needed to enable a new level of understanding about the behavior of advanced materials under relevant operating conditions. Advances in spectroscopy and microscopy, as well as new combinations of *in-situ* and *ex-situ* methods, enables real time measurements using a variety of synchrotron and laboratory-based capabilities. In particular, the utilization of novel ambient pressure x-ray, electron and vibration spectroscopy techniques to understand oxidation and catalytic reactions on surfaces in energy and environmental science studies plays a crucial role in the advancement of science and technology. New developments in electron microscopy, including aberration-corrected microscopes coupled with environmental cells, allow unique *in-situ* experiments to study structure/property & stimuli/response relationships, and related dynamic processes at or near atomic/molecular level. Finally, neutron tomography and radiography allow us to look internally at devices and systems under operation.

Chongmin Wang
Niels de Jonge
Rafal E. Dunin-Borkowski
Artur Braun
Jinghua Guo
Randall E. Winans
Helmut Schober

July 2010

ACKNOWLEDGMENTS

The organizers would like to thank all of the speakers (including invited speakers), the session chairs, and the symposium assistants (Manjula Nandasiri of Michigan State University for Symposium V, and Qianli Chen of ETH Zürich and Empa for Symposium W) for making both symposia a success. The organizers are grateful to the MRS editorial staff for their diligent work in assembling this volume.

Financial support for Symposium V was provided by:

- EA Fischione Instruments, Inc.
- Hitachi High Technologies America, Inc.
- IBM Thomas J. Watson Research Center
- JEOL USA, Inc.

Financial support for symposium W was provided by:

- Empa — Swiss Federal Laboratories for Materials Science and Technology
- Huber Diffraktionstechnik GmbH & Co. KG
- National Renewable Energy Laboratory
- SPECS Surface Nano Analysis GmbH
- SwissNeutronics — Neutron Optical Components & Instruments

MATERIALS RESEARCH SOCIETY SYMPOSIUM PROCEEDINGS

MATERIALS RESEARCH SOCIETY SYMPOSIUM PROCEEDINGS

Prior Materials Research Society Symposium Proceedings available by contacting Materials Research Society

In-Situ Nanoscale Deformation

Mater. Res. Soc. Symp. Proc. Vol. 1262 © 2010 Materials Research Society
1262-V01-04

In situ TEM Straining of Nanograined Al Films Strengthened with Al$_2$O$_3$ Nanoparticles

K. Hattar[1], B. G. Clark[1], J. A. Knapp[1], D. M. Follstaedt[1], and I. M. Robertson[2]
[1] Sandia National Laboratories, Physical, Chemical, & Nano Sciences Center, PO Box 5800 Albuquerque, NM 87185, U.S.A.
[2] University of Illinois at Urbana-Champaign, Department of Materials Science and Engineering, 1304 W. Green Street, Urbana, IL 61801, U.S.A.

ABSTRACT

Growing interest in nanomaterials has raised many questions regarding the operating mechanisms active during the deformation and failure of nanoscale materials. To address this, a simple, effective *in situ* TEM straining technique was developed that provides direct detailed observations of the active deformation mechanisms at a length scale relevant to most nanomaterials. The capabilities of this new straining structure are highlighted with initial results in pulsed laser deposited (PLD) Al-Al$_2$O$_3$ thin films of uniform thickness. The Al-Al$_2$O$_3$ system was chosen for investigation, as the grain size can be tailored via deposition and annealing conditions and the active mechanisms in the binary system can be compared to previous studies in PLD Ni and evaporated Al films. PLD Al-Al$_2$O$_3$ free-standing films of various oxide concentrations and different thermal histories were produced and characterized in terms of average grain and particle sizes. Preliminary *in situ* TEM straining experiments show intergranular failure for films with 5 vol% Al$_2$O$_3$. Further work is in progress to explore and understand the active deformation and failure mechanisms, as well as the dependence of mechanisms on processing routes.

INTRODUCTION

There has been significant interest in nanograined metals and their deformation and failure mechanisms for several decades. Over this time, the processing and mechanical testing of these nanograined structures have improved significantly. Many interesting properties have been reported and attributed to a variety of mechanisms [1, 2]. For example, Wang et al. has reported that with the proper control of rolling at liquid nitrogen temperature, followed by low temperature annealing, a metal that has both high strength and large ductility can be formed. This combination of properties is not typical of either nanograined or coarse grained metals [3]. In order to understand the active mechanisms that permit this unique combination of properties, detailed *in situ* TEM experiments must be done to investigate the effect of microstructural variables within nanograined films.

The development of *in situ* TEM techniques for straining of free-standing thin films and simultaneous observations of the active deformation and failure mechanisms has progressed significantly over the last decade. A variety of *in situ* TEM straining structures used to apply mechanical load to nanostructured metals has been developed. Each of these structures provide benefits and limitations that should be considered when developing an *in situ* experiment to elucidate certain mechanisms [4, 5]. This study will emphasize a custom-made straining structure that was specifically developed for this study based on previous straining structures in the literature [6, 7].

Previous studies using a similar pulsed-laser deposition (PLD) technique coupled with *in situ* TEM straining structures revealed interesting results in PLD Ni thin films of various microstructures [8]. In post-mortem analysis, the straining of nanograined, ultra-fine grained, and bimodal PLD Ni films of nominally 80 nm thickness demonstrated three different, corresponding fracture surfaces: The nanograined films showed a smooth fracture surface with little plasticity, the ultra-fine grained films showed a jagged fracture surface with localized plasticity, and the films with a bimodal grain size distribution showed a mixture of the two. In the latter case, the fracture surface was smooth when it progressed through nanograined regions, but showed necking, dislocation pile-ups, and twinning when the fracture path encountered large grains. Additional observations during *in situ* straining showed further details of the active mechanisms during deformation [8], and thus validated the use of simple TEM straining structures for understanding failure mechanisms.

In a separate study, ultra-fine grained Al sputter deposited and electron-beam evaporated thin films were found to be columnar in nature with deformation and failure dominated by grain boundary grooving. In these films, which were deposited on Si and made free-standing by a series of microfabrication steps, the surface was observed to have significant surface roughness. *In situ* TEM straining experiments of the ultra-fine grained Al films resulted in intergranular failure with limited to no plasticity throughout the film. All of the observed plasticity occurred directly ahead of the crack tip. Large grains in these films acted to hinder crack progression and showed limited dislocation activity and twinning. The intergranular failure was inhibited by increasing the film thickness, which decreased the percentage of stress-riser to film thickness ratio [8]. The insight gained from *in situ* TEM straining experiments of high purity nanograined PLD Ni and ultra-fine grained Al samples provides two experimental parameters to compare with this investigation into nanograined Al-Al$_2$O$_3$ films in which greater complexity in the active mechanisms is expected.

EXPERIMENT

Nanograined Al films strengthened with Al$_2$O$_3$ precipitates were prepared using PLD. Using two alternating sources, mixed films of Al and Al$_2$O$_3$ were deposited onto polished NaCl substrates with final compositions of 1, 5, and 10 vol% Al$_2$O$_3$. For each volume percent, three deposition routes were used varying the number of deposition cycles (5, 10, or 20). In total 9 films were deposited, each with a final thickness of 100 nm. Based on past experiments, the films are assumed to show some columnar nature and be predominantly [111] in texture with some surface roughness. To coarsen the Al$_2$O$_3$ particles that were precipitated during deposition, and to additionally grow the grain structure to facilitate *in situ* TEM observations, the films were annealed in a vacuum furnace. Anneals took place for 1 hour at 300, 400 and 500 °C at 1 x 10-5 Torr.

To prepare free-standing thin films for straining, the following preparation steps were followed, see Figure 1. First, sacrificial straining structures were prepared by cutting thin sheets of 301 stainless steel into 2.4 x 11 mm rectangles, drilling holes at each end, and using a diamond saw to cut a small slit from edge-to-center at the middle of each rectangle, Figure 1a. Next, the above described films of PLD Al-Al$_2$O$_3$ deposited onto NaCl substrates were cleaved into rectangular blocks, approximately 1.5 x 5 mm in cross section. Each NaCl block with PLD film attached was glued using M-bond, film side down, to a straining structure, Figure 1b. After the M-bond had cured at room temperature, the NaCl substrate was dissolved in a bath of deionized (DI) water, Figure 1c. A series of steps were then taken to remove the dissolved NaCl from the DI water by pipetting out most of the fluid, refilling with DI water, and repeating. In the final steps, ethanol was added to the solution, instead of DI water, to reduce the surface tension of the fluid and prevent rupture of the free-standing thin film upon drying. After these steps, the film was removed from the solution with tweezers and held in air to dry prior to straining in the TEM. An actual straining structure with an 80 nm-thick PLD film arrowed is shown in Figure 1d.

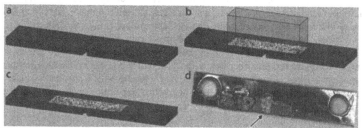

Figure 1. a) Sacrificial straining structure; b) NaCl with deposited film attached to straining structure; c) straining structure with the NaCl removed, leaving the film adhered to the structure; and d) an example of an actual structure with an 80 nm-thick PLD Ni film attached to a straining structure.

DISCUSSION

The as-deposited microstructures for Al-Al$_2$O$_3$ films were characterized using plan-view bright-field TEM over a range of vol.% Al$_2$O$_3$ and number of deposition cycles. The films were all nanograined, but did not all contain identifiable particles. No particles were seen in the case of the 1 vol.% films, and the particles are sparse in the 5 and 10 vol.% films.

In order to develop microstructures with observable Al$_2$O$_3$ particles and larger grain sizes to facilitate TEM imaging and analysis, a series of annealing experiments were performed for each film: 300, 400, and 500 °C for 1 hour. An example comparison of a 1 vol.% Al-Al$_2$O$_3$ film deposited using 10 deposition cycles is shown in Figure 2 for (a) an as-deposited and (b) a post-annealing microstructure. Here it is clear that the annealing experiment both increased the number of observable precipitated Al$_2$O$_3$ particles and the average grain size of the film. The average grain size in the as-depostied film, Figure 2a, was 31 nm with no distinguishable Al$_2$O$_3$ particles. After annealing, Figure 2b, the average grain size was found to be 110 nm and the average particle size 9 nm. Over the range of the 18 post-annealing films analyzed, the average particle size ranged from 3 to 12 nm and the average grain size from 30 to 110 nm.

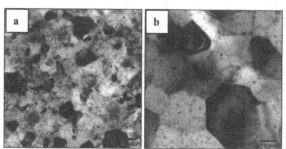

Figure 2. Examples of microstructures in a 1 vol.% Al-Al$_2$O$_3$ film deposited using 10 deposition cycles. (a) As deposited microstructure, and (b) microstructure after annealing 1 hour at 500 °C showing increase in grain size and development of Al$_2$O$_3$ particles.

Shown in Figure 3 is an example matrix of microstructures showing the relationship between vol.% Al_2O_3 and annealing temperature. All films were deposited using 5 layer pairs. This figure shows a few key features: In films with lower vol.% Al_2O_3, the final grain sizes were larger (compare, for example, g, h & i). This is likely due to fewer precipitates present, and thus fewer obstacles pinning the movement of grain boundaries during annealing. Additionally, anneals at higher temperatures led to larger grain sizes, as well (compare, for example, a, d, & g). This is due to increased mobility of grain sizes, as temperature is increased, resulting in faster grain growth rates. Microvoids are observed to form at grain boundaries and triple points in 1 vol% films annealed at 500 °C for one hour. This is associated with the rapid growth at 0.83 of the melting temperature and no Zener drag effect and will serve, as the limit for thermal processing of these films. Finally, in all 9 films, Al_2O_3 precipitates were observed after annealing. These observations for films deposited with 5 deposition cycles are consistent with films deposited using 10 and 20 deposition cycles.

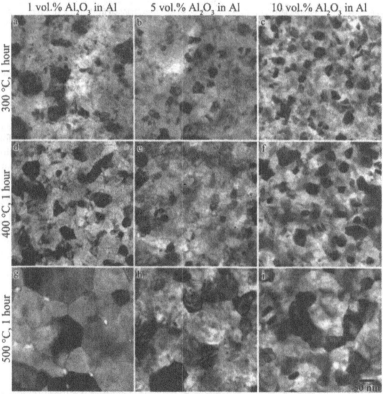

Figure 3. Post-annealing microstructures formed in Al-Al_2O_3 films deposited using 5 layers pairs showing relationship between vol.% Al_2O_3 and annealing temperature.

6

Following the development of these microstructures and preparation of free-standing straining samples, preliminary *in situ* TEM straining experiments were performed. Figure 4 shows a 1 vol.% Al-Al$_2$O$_3$ film, deposited using 10 layer pairs and annealed 1 hour at 400 °C during *in situ* TEM straining. Although the crack path was intergranular, which is associated with brittle fracture, microcracking ahead of the crack tip and dislocation emission during crack opening indicate the presence of toughening mechanisms in the film.

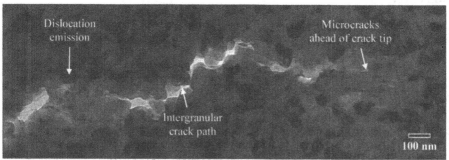

Figure 4. Crack formed in a 1 vol.% Al-Al$_2$O$_3$ film (deposited using 10 layer pairs, annealed 1 hour at 400 °C) during *in situ* TEM straining.

Future work in the Al-Al$_2$O$_3$ system will include additional annealing experiments and *in situ* TEM straining experiments to understand the link between structure and mechanical properties in nanoparticle-strengthened, nanograined thin films over a range of controlled microstructures. The aim will be to develop stable microstructures that enhance the ductility and toughness of these films and determine the underlying mechanisms that operate in these microstructures resulting in enhanced properties. In addition to the investigation into Al-Al$_2$O$_3$, future work will include understanding the effects of particle type (e.g. soft vs. hard, coherent vs. incoherent) and the behavior of more complex systems (e.g. multiple constituents, multiple particles/phases, solute) on the active mechanisms and resulting properties.

CONCLUSIONS

In situ TEM investigation into the deformation and failure mechanisms in nanograined and ultra-fine grained metals was pursued using a custom-built straining structure. Details of an initial investigation into the deformation and failure mechanisms in free-standing PLD Al-Al$_2$O$_3$ films were presented. Thermal processing was used to tailor the Al$_2$O$_3$ particle size from 3 nm to 12 nm and the Al grain size from 30 nm to 110 nm. The initial results suggest that failure in the 5 vol% Al$_2$O$_3$ in Al films is predominantly intergranular with limited signs of plasticity. Further work is underway to understand the effect of various particle sizes and interfaces in nanograined alloys.

ACKNOWLEDGMENTS

The authors would like to thank Dr. L.N. Brewer, Prof. P. Sofronis, C.J. Powell, J. Kacher, and J.A. Fenske for many fruitful conversations, and M. Moran for assistance in analysis and sample preparation. This work was in part sponsored by Department of Energy BES through award DE-FG02-07ER46443 (IMR). This work is supported by the Division of Materials Science and Engineering, Office of Basic Energy Sciences, U.S. Department of Energy. Sandia National Laboratories is a multi-program laboratory operated by Sandia Corporation, a wholly owned subsidiary of Lockheed Martin company, for the U.S. Department of Energy's National Nuclear Security Administration under contract DE-AC04-94AL85000.

REFERENCES

[1]   M. A. Meyers, A. Mishra and D. J. Benson, Progress in Materials Science 51 (2006) 427.
[2]   H. Conrad, Metallurgical and Materials Transactions A: Physical Metallurgy and Materials Science 35 A (2004) 2681.
[3]   Y. Wang, M. Chen, F. Zhou and E. Ma, Nature 419 (2002) 912.
[4]   K. Hattar and D. M. F. J.H. Han, S. J. Hearne, T. A. Saif, I. M. Robertson, in Fall MRS Conference 2005, edited by I. M. R. Paulo J. Ferreira, Gerhard Dehm, Hiroyasu Saka (MRS Publication, Boston, MA, 2005).
[5]   J. A. Fenske, K. Hattar, M. Briceno and I. M. Robertson, Development of in-situ TEM straining structures, (University of Illinois, 2010).
[6]   R. C. Hugo, H. Kung, J. R. Weertman, R. Mitra, J. A. Knapp and D. M. Follstaedt, Acta Materialia 51 (2003) 1937.
[7]   R. D. Field and P. A. Papin, Ultramicroscopy 102 (2004) 23.
[8]   K. Hattar, Thermal and mechanical stability of nanograined FCC metals, in "Materials Science and Engineering" (University of Illinois, Urbana, 2009) p. 237.

Poster Session: *In-Situ* **TEM**

Mater. Res. Soc. Symp. Proc. Vol. 1262 © 2010 Materials Research Society   1262-V05-01

Direct Observations of Relaxation of Strained Si/SiGe/Si on Insulator

Tongda Ma and Hailing Tu
General Research Institute for Nonferrous Metals, No.2 Xinjiekouwai Street, Beijing, 100088, P. R. China

ABSTRACT

Microstructural evolution is directly observed when the cross-sectional film specimen of Si/SiGe/Si on insulator (Si/SiGe/SOI) is heated from room temperature (R.T., 291 K) up to 1113 K in high voltage transmission electron microscope (HVEM). The misfit dislocation at the lower interface of the SiGe layer begins to extend downwards even at 913 K. The lower interface takes the lead in roughening against the upper interface of the SiGe layer. The roughened interface is ascribed to elastic relaxation. As misfit strain is partially transferred to SOI top Si layer and misfit dislocation is prolonged at the lower interface, the roughened interface turns smooth again. Thereafter, the misfit dislocations are introduced into the upper roughened interface of the SiGe layer to release the increased misfit strain. It is suggested that the microscopic relaxation of the SiGe layer is related to dislocation behavior and strain transfer.

INTRODUCTION

The past several years have witnessed rapid growth in the study of strained Si on insulator due to its potential ability to improve the performance of very large scale integrated circuits [1]. In order to fabricate tensile strained Si capping layer, some creative methods have been reported on the basis of different mechanisms [2-5]. Most of mechanisms essentially depend on the compliancy of buried oxide (BOX) under high temperature. To evaluate the strain engineering, transmission electron microscopy (TEM) and high resolution x-ray diffractometry (HRXRD) are usually adopted to characterize the strain transfer and defect behavior [4, 5]. As a useful complementarity, in-situ TEM is performed to trace strain relaxation of the SiGe or Si/SiGe on bulk Si [6-8]. However, few in-situ TEM works have ever been focused on Si/SiGe/SOI by now.

It is demonstrated that strained Si/SiGe/SOI can be achieved by the successive deposition of the Si buffer, the SiGe layer and the Si capping layer on a bonded SOI wafer and the following in-situ thermal treatment [9]. TEM, triple-axis x-ray diffractometer (TAD), and secondary ion mass spectrometer (SIMS) have been employed to investigate defects, strain, and interdiffusion within Si/SiGe/Si on insulator. In this work, HVEM is used to observe microstructral evolution of Si/SiGe/SOI during in-situ annealing in order to further clarify the mechanism of strain relaxation.

EXPERIMENTAL DETAILS

Si/SiGe/SOI is grown in the ultra-high vacuum chemical vapor deposition (UHV/CVD) system. The starting bonded SOI substrate is chemically cleaned for 15 minutes in a boiled solution of H_2SO_4:H_2O_2 (4:1), then rinsed in de-ionized water for at least 10 minutes. Native oxide on the wafer surface is etched in 10% HF solution for 30 seconds and then ferried into the loading chamber. When the vacuum reaches 10^{-5} Pa, the wafer is put into the growth chamber. The growth chamber is pumped with a 1000 l/s turbo-molecular pump and a base pressure of 5×10^{-7} Pa is obtained. The chamber pressure is maintained below 0.13 Pa during the epitaxial process. As soon as the base vacuum is approached, the temperature is raised to 1023–1073 K at a high ramp-up rate. After the wafer is heated in the vacuum for 5 minutes, the temperature descends to the growth temperature, for Si, 873 K and for SiGe, 823 K. Pure SiH_4 and 15% GeH_4 diluted in H_2 are used for SiGe epitaxial growth. The flow rates of SiH_4 and GeH_4 are 10 and 2 sccm, respectively during the growth. The final structure from top to bottom is composed of 20 nm-thick Si capping layer, 40 nm-thick $Si_{0.8}Ge_{0.2}$ alloy layer, 15 nm-thick Si buffer layer, and the bonded SOI substrate (with 85 nm-thick top Si layer).

Direct observations of strain relaxation of Si/SiGe/SOI are performed using JEM-1000. 1M volts are selected as the work voltage. Such high work voltage is very suitable for imaging dislocations in thick films as similar as the bulk materials. The heating rate is 10~20 K/minute. Measurement precision for temperature is 20 K. The cross-sectional film specimen is prepared by mechanical thinning to about 50 μm, then dimpling to 10 μm, followed by Ar ion milling at 3.5 kV to electron transparency. The in-situ observation in HVTEM is done from room temperature (R.T., 291 K) up to 1113 K. After thermal treatment, the specimen is cooled down to R.T. along with the heating stage. The microstructural evolvement is recorded on electron sensitive films.

DISCUSSION

Misfit strain due to the mismatch between the SiGe layer and SOI top Si layer can be partially released by formation of misfit dislocations during the post-growth thermal treatment (1023 K, 30 minutes) (Figure 1). This has also been proved by former research [9, 10].

Figure 1. Dislocation A forms at the lower interface of the SiGe layer and the interface between the SiGe layer and Si layers is clearly visible. (The scale bar corresponds to 35nm in all the figures.)

During in-situ annealing in HVEM, no obvious changes happen to the cross-sectional film specimen till heated at 913 K for 20 minutes (Figure 2).

Figure 2. The cross-sectional image of Si/SiGe/SOI is taken at 20th minute after the film specimen is heated to 913K in HVEM. Dislocation A keeps unchanged by that time.

However, as in-situ annealing lasts, strain relaxation is activated at lower temperature than that of post-growth thermal treatment. This should be attributed to in-situ relaxation of an electron transparent cross-section different from that of a large area film. Specially, the microscopic relaxation is easier than macroscopic relaxation due to the small width-to-thickness ratio of the films [11]. Dislocation B forms on the surface and slides cross the upper interface of SiGe layer for little misfit strain existing at the upper interface. Dislocation A is responsible for plastic relaxation and elastic relaxation makes the lower interface of the SiGe layer roughened.

Figure 3. Serious variation happens at 44 minutes later when the film specimen is heated to 913K in HVEM. The lower interface turns rough. At the same time, Dislocation A continues to extend downwards and Dislocation B generates at the surface and slides towards the lower interface of the SiGe layer.

Compliancy has been demonstrated on mesas which would suggest it might be possible on a less than 10 micron wide sliver that is heated in the TEM [12]. The decreased roughness at the lower interface of the SiGe layer indicates that misfit strain may be transferred to the SOI top Si layer while the BOX becomes compliant (Figure 4). Meanwhile, the upper interface gets rough due to strain relaxation of the SiGe layer.

Figure 4. The cross-sectional image of Si/SiGe/SOI is taken at 12 minutes later when the film specimen is heated to 1033K. Dislocation A further approaches to the interface between SOI top Si layer and BOX. Dislocation B reaches to the lower interface of the SiGe layer. The roughness of the lower interface decreases in contrast to the upper interface of the SiGe layer.

The threading part of Dislocation B moves out of the film specimen and Dislocation A slides down to the interface between SOI top Si layer and BOX (Figure 5). The misfit dislocations of Dislocation A and B prolong at the lower interface of the SiGe layer which finally result into more strain relaxation. Once the elastic deformation cannot dissipate the increasing mismatch caused by the relaxation at the lower interface of the SiGe layer, misfit dislocations are introduced into the upper interface (not shown here).

Figure 5. The cross-sectional image of Si/SiGe/SOI is taken at 1113K. It shows Dislocation A arrives at the interface between SOI top Si layer and BOX. The threading part of Dislocation B shifts out of the film specimen.

CONCLUSSIONS

The cross-sectional film specimen of Si/SiGe/SOI is in-situ heated from R.T. to 1113 K in HVEM. The strain relaxation of the SiGe layer is observed at the lower temperature than that of post-growth thermal treatment due to the size effect of film specimen. The prolonged misfit dislocations and strain transfer are two major reasons for mircoscopic relaxation at the lower interface of the SiGe layer. This triggers off the strain relaxation at the upper interface of the SiGe layer.

ACKNOWLEDGEMENTS

This work is sponsored by the contract No. 50502008 and No. 50872013 from National Natural Science Foundation of China.

REFERENCES

1. K. Rim, J. L. Hoyt, and J. F. Gibbons, IEEE Trans. Electron Devices 47, 1406 (2000)
2. A. R. Powell, S. S. Iyer and F. K. Legoues, Appl. Phys. Lett. 64, 1856 (1994)
3. Y. H. Luo, J. L. Liu, G. Jin, J. Wan, K. L. Wang, Appl. Phys. A 74, 699 (2002)
4. P. M. Mooney, G. M. Cohen, J. O. Chu, and C. E. Murray, Appl. Phys. Lett. 84, 1093 (2004)

5. E. M. Rehder, C. K. Inoki, T. S. Kuan, T. F. Kuech, J. Appl. Phys. 94, 7892 (2003)
6. R. Hull, J. C. Bean, D.J.Werder, and R.E.Leibenguth, Appl. Phys. Lett. 52, 1605 (1988)
7. E. A. Stach, R. Hull, J. C. Bean, K. S. Jones, and A. Nejim, Microscopy AND Microanalysis 4, 294 (1998)
8. Yu. B. Bolkhovityanov, A. S. Deryabin, A. K. Gutakovskii, M. A. Revenko, and L. V. Sokolovb, Appl. Phys. Lett. 85, 6140 (2004)
9. T. Ma, H. Tu, G. Hu, B. Shao, A. Liu, J. Crys. Growth 289, 489 (2006)
10. T. Ma, H. Tu, G. Hu, B. Shao, and A. Liu, Appl. Surf. Sci. 253, 124 (2006)
11. G. Kästner and U. Gösele, J. Appl. Phys. 88, 4048 (2000)
12. M. M. Roberts, L. J. Klein, D. E. Savage, K. A. Slinker, M. Friesen, G. Celler, M. A. Eriksson and M. G. Lagally, Nature Materials 5, 388 (2006)

Mater. Res. Soc. Symp. Proc. Vol. 1262 © 2010 Materials Research Society

Three-Dimensional Structure of Twinned and Zigzagged One-Dimensional Nanostructures Using Electron Tomography

Han Sung Kim, Yoon Myung, Yong Jae Cho, Dong Myung Jang, Chan Soo Jung, Jae-Pyoung Ahn, and Jeunghee Park
Department of Chemistry, Korea University, Jochiwon 339-700, Korea

ABSTRACT

Electron tomography and high-resolution transmission electron microscopy were used to characterize the unique 3-dimensional (3D) structures of twinned Zn_3P_2 (tetragonal) and InAs (zinc blende) nanowires synthesized by the vapor transport method. The Zn_3P_2 nanowires adopt a unique superlattice structure that consists of twinned octahedral slice segments having alternating orientations along the axial [111] direction of a pseudocubic unit cell. The apices of the octahedral slice segment are indexed as six equivalent <112> directions at the [111] zone axis. At each 30 degrees turn, the straight and zigzagged morphologies appear repeatedly at the <112> and <011> zone axes, respectively. The 3D structure of the twinned Zn_3P_2 nanowires is virtually the same as that of the twinned InAs nanowires. In addition, we analyzed the 3D structure of zigzagged CdO (rock salt) nanowires and found that they include hexahedral segments, whose six apices are matched to the <011> directions, linked along the [111] axial direction. We also analyzed the unique 3D structure of rutile TiO_2 (tetragonal) nanobelts; at each 90 degree turn, the straight morphology appears repeatedly, while the in-between twisted form appears at the [011] zone axis. We suggest that the TiO_2 nanobelts consist of twinned octahedral slices whose six apices are indexed by the <011>/<001> directions with the axial [010] direction.

INTRODUCTION

One-dimensional (1D) nanostructures have attracted considerable attention due to their potential use as building blocks for assembling active and integrated nanosystems.[1] Recently, the interest in twinned superlattice 1D nanostructures that have twin planes at a constant spacing has been steadily increasing, owing to their attractive morphology and electrical/optical properties.[2-23] Since a twin boundary can act as a natural potential well for electrons, the discontinuous electron wavefunction leads to a reduction in the mobility of the charge carriers. For twinned zinc blende (ZB)/wurtzite (WZ) InP heterostructure nanowires (NWs), observed the excitation power-dependent blue-shift of the photoluminescence and explained it in terms of the staggered band alignment and concomitant diagonal transition between the localized electron/hole states.[15] Calculations predicted that a constant spacing between the rotational twins would induce a direct bandgap in normally indirect bandgap semiconductors, such as group IV (Si, Ge) and III-V (GaAs) materials.[23] Therefore, the controlled formation of a twinning structure in relevant semiconductors could have a significant impact on optically active band-structure engineering. In many semiconductor and noble metal NWs with a cubic structure (e.g., GaP, InP, InAs, GaAs, ZnS, ZnS, ZnTe, Si, SiC, B_4C, Cr_2O_3, Zn_2SnO_4, Zn_2TiO_4, Ag, Au), (111) twin planes are commonly found.[2-22] However, only a few twinned structures have been reported for other crystal other crystal structures.[9,12,14,18] Electron tomography, which is a method of reconstructing the 3D morphologies from a series of two-dimensional (2D) transmission electron microscopy (TEM) images or projections, has been successfully applied to analyze the morphology of various 1D nanostructures.[24-29] Herein, we employed both electron tomography and highresolution TEM images to study the 3D structure of various twinned and zigzagged 1D nanostructures. Their 3D reconstruction images, acquired from a series of 2D projections, were obtained by high-angle annular dark field (HAADF) scanning TEM (STEM).[30] Zn_3P_2 belongs to a unique tetragonal system, whose lattice constant, c/a, ratio (1.414) leads to its having a pseudo face-centered cubic (fcc) symmetry.[31,32] Zn_3P_2 NWs (including twinned NWs), nanobelts (NBs), and nanotubes were usually synthesized using vapor phase transport, and their electronic and optoelectronic properties were also indentified.[12,33-35] The present work demonstrates that tetragonal Zn_3P_2 NWs have virtually the same twinned superlattice structure as that of ZB InAs NWs. In addition, the 3D structure of rock-salt cubic CdO NWs was analyzed to show their unique zigzagged structures, which are worth comparing with the twinned structures. Furthermore, we studied the novel 3D structure of zigzagged rutile tetragonal TiO_2 NBs, which appear to have another unique twinned structure model.

EXPERIMENTAL SECTION

(1) **Materials :** The Zn_3P_2, InAs, and CdO NWs were all synthesized by the vapor transport method utilizing the vapor-liquid-solid growth mechanism. To synthesize the Zn_3P_2 NWs, a mixture of Zn (Aldrich, 99.999 %) and InP (Aldrich, 99.99%) powders was placed in a ceramic boat, in order to generate Zn and P vapors, and 3 nm-thick Au filmdeposited Si substrates were placed 20-cm apart from the source. An argon flow at 50~200 sccm was introduced into the reactor tube under ambient pressure, followed by heating to 800-850 ℃. The growth reaction temperature at the substrates was maintained at 600 ℃ for 60 min. The InAs NWs were synthesized using the thermal evaporation of InAs (Aldrich, 99.99 %) powders at 900 ℃ and 3 Torr. The substrates were Au film (3 nm thickness)-deposited Si substrates. An argon flow at 100 sccm was maintained during the whole growth process and the reaction time was 30-60 min. To synthesize the CdO NWs, a mixture of Cd (Aldrich, 99.999 %) and CdO (Aldrich, 99.99%) powders was placed in a ceramic boat, in order to generate Cd vapor, and 3 nm-thick Au film-deposited Si substrates were placed 20-cm apart from the source. An argon flow at 200 sccm was introduced into the reactor tube under ambient pressure, followed by heating to 1100 ℃. The growth reaction at the substrates was maintained at 900 ℃ for 60-90 min. The TiO_2 NBs were synthesized by the thermal oxidation method, following the solid-liquid-solid (SLS) growth mechanism. Ultrasonically cleaned Ti foils ($10 \times 10 \times 0.25$ mm) with a purity of 99.7 % (Aldrich) were coated with a 3 nm-thick Au film and used as both the reagent and substrates for the growth of the TiO_2 NBs. The Ti foil was placed on the top of a quartz boat, located inside a quartz tube reactor. The temperature of the Ti foil was set to 800 °C. A flow of oxygen (> 99.999 %) with a rate of 20-50 sccm was introduced only for the reaction time of 1 h. Then the Ti foil was covered homogeneously with the white-colored NW array.

(2) **Characterizations :** The products were analyzed by scanning electron microscopy (SEM, Hitachi S-4700), field-emission transmission electron microscopy (TEM, FEI Co. TECNAI F20 G2 200 kV and Jeol JEM 2100F), high-voltage TEM (HVEM, Jeol JEM ARM 1300S, 1.25 MV), electron diffraction (ED), and energy-dispersive X-ray fluorescence spectroscopy (EDX). High-resolution X-ray diffraction (XRD) patterns were obtained using the 8C2 and 3C2 beam lines of the Pohang Light Source (PLS) with monochromatic radiation (λ = 1.54520 Å).

(3) **Electron Tomography :** 3D electron tomography was performed using a STEM (FEI Co., TECNAI F20 G2 200 kV), with a tilt holder (Dual Orientation Tomography Holder 927, Gatan Co.) and a Fischione model 3000 HAADF detector operated at 200 kV. A series of 130 HAADF-TEM images was collected from +75° to −75° in 1.5° steps under a nominal.

18

RESULTS AND DISCUSSION

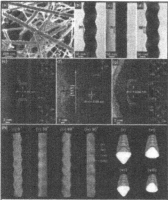

Figure 1. Images of the zigzagged Zn_3P_2 NWs obtained through SEM, TEM and Electron Tomography.

Figure 1: (a) SEM micrograph of high-density Zn_3P_2 NWs homogeneously grown on the substrate. (b) TEM image reveals the zigzagged morphology of the Zn_3P_2 NW having a zigzagged period of 140 nm. Consecutive 30° tilt around the wire axis changes the morphology to (c) straight and (d) zigzagged. Lattice-resolved TEM images of the (e) 0, (f) 30, and (g) 60°-turn morphologies. The insets show the corresponding FFT ED patterns at the $[01-1]_c$ zone axis for the 0° turn, $[11-2]_c$ for the 30° turn, and $[10-1]_c$ for the 60° turn. The $[111]_c$ axial direction remains the same for each tilt. The distance between the adjacent $(111)_c$ planes of the pseudo cubic unit cell is 6.6 Å. At the $[01-1]_c$ zone axis, two segments share the $[111]_c/[-1-1-1]_c$ spots. The (i) and (iii) ED patterns correspond to those of the twin segments and (ii) the ED pattern corresponds to their twin plane region. (h) Images of the Zn_3P_2 NW obtained through tomographic 3D reconstruction; (i)-(iv) correspond to a series of 0°, 30°, 60°, and 90° tilt around the axial direction,

respectively; (v)-(viii) images show the sliced views along the NWs (as marked in (iv)), having a triangular cross section.

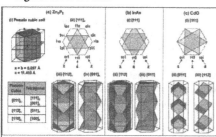

Figure 2. Schematic model constructed for the zigzagged Zn_3P_2, InAs, and CdO NWs.

Figure 2: (a) Zn_3P_2 NWs; (i) pseudo cubic unit (fcc) cell in which the parameters a of the tetragonal cell are directed along the diagonals of the faces of the elementary cube, and the parameter c is equal to the doubled edge of the cube. The table shows the index correlation between tetragonal and cubic unit cells. (ii) Schematic model constructed for the twinned Zn_3P_2 NWs at the $[111]_c$ zone axis; the twin octahedral slice blocks have six equivalent $<112>_c$ apices. The six side facets of the blocks would be enclosed by the $\pm(11-2)$, $\pm(1-21)$,

and $\pm(2-1-1)$ surfaces. Schemes (iii) and (iv) correspond to the views for the $[112]_c$ and $[011]_c$ zone axes, respectively, showing the straight (symbolized as "s" in (ii)) and zigzagged (symbolized as "z" in (ii)) morphologies. (b) Schematic model constructed for the twinned InAs NWs. (i) At the [111] zone axis, the twin octahedral slice blocks have six equivalent <112> apices. The side facet of the blocks is enclosed by the {112} planes. Schemes (iii) and (iv) correspond to the views at the <112> and <011> zone axes, respectively, showing the straight and zigzagged morphologies. (c) Schematic model constructed for the CdO NWs. (i) The apices of the hexahedral segments have six equivalent <011> directions at the [111] zone axis: [01-1], [0-11], [10-1], [-101], [-110], [1-10], and the axial direction is [111]. (ii) At the <011> zone axes, the segments appear as hexagons, (iii) while at the <112> zone axes, their shape becomes rhombohedral.

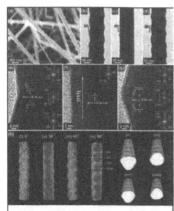

Figure 3: (a) SEM micrograph of high-density InAs NWs homogeneously grown on the substrate. (b) TEM image reveals the zigzagged morphology of the InAs NW (diameter = ~100 nm) with a period of 80 nm. (c) TEM image for its 30° tilt, showing the straight morphology. (d) Another 30° sequential tilt leads to the production of a zigzagged shape. Lattice-resolved TEM images of the (e) 0, (f) 30, and (g) 60°-turn morphologies and their corresponding FFT ED patterns are shown in the insets. The zone axis is indexed as [01-1] for the 0° turn, [11-2] for the 30° turn, and [10-1] for the 60° turn. The distance between the adjacent (111) planes is 3.5 Å. At the [01-1] and [10-1] zone axes, ED patterns (i) and (iii) correspond to those of the twin segments, and pattern (ii) corresponds to the twin plane region. (h) Tomographic 3D reconstruction images; (i)-(iv) images for the 30°sequential turns. The sliced views along the NW (as marked in (iv), (v)-(viii)), reveal the hexagonal cross-section.

Figure 3. Images of the zigzagged InAs NWs obtained through SEM, TEM and Electron Tomography.

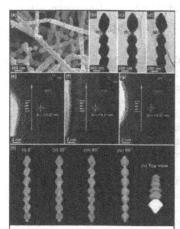

Figure 4: (a) SEM image showing high-density chain-like CdO NWs grown on the substrates. (b) TEM image reveals that the NW uniformly consists of linked rhombohedral segments. (c) As it is tilted by 30°, the shape of the segments changes to hexagonal. (d) As it is further tilted by 30°, the shape of the segments returns to rhombohedral. Lattice-resolved TEM images for the (e) 0°, (f) 30°, and (g) 60°-tilt. The distance between the adjacent (111) planes is 2.7 Å. The zone axes of the corresponding SAED patterns are [01-1], [11-2], and [10-1], for the 0°, 30°, and 60°-tilt (insets), respectively. The wire axis is [111]. (f) Tomographic 3D reconstruction images: (i)-(iv) correspond to 0, 30, 60, and 90°-turns, showing the shape change of the segments from rhombohedral to hexagonal and vice versa, as the NW is tilted by 30°. The top-view (v) reveals its cross-section.

Figure 4. Images of the zigzagged CdO NWs obtained through SEM, TEM and Electron Tomography.

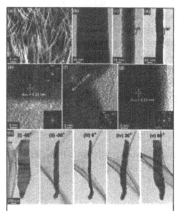

Figure 5. Images of the zigzagged TiO$_2$ NWs obtained through SEM, TEM and Electron Tomography.

Figure 5: (a) SEM image of the high-density TiO$_2$ NBs on the Ti substrates. (b) TEM image of the NBs (width = 300 nm). TEM images for their (c) 40° tilt, showing the zigzagged structure, and (d) 90° tilt, producing the straight morphology. Latticeresolved images of the (e) 0, (f) 40, and (g) 90°-turn morphologies and their corresponding FFT ED patterns (insets). The zone axis is indexed as [100] for the 0° turn, [011] for the 40° turn, and [001] for the 90° turn. The [010] axial direction remains the same for each tilt. The distance between the adjacent (100) planes is 2.3 Å. (h) A series of TEM images of another NB showing the morphology change for the -60, -30, 0, 30, and 60 °-tilt.

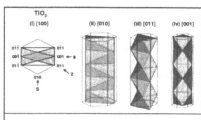

Figure 6. Schematic model constructed for the zigzagged TiO$_2$ NBs.

Figure 6: Schematic model constructed for the TiO$_2$ NB at the (i) [100], (ii) [010], (ii) [011], and [001] zone axes. (i) The octahedral blocks have six apices at the <011>/<001>, i.e., [0-1-1], [0-11], [01-1], [011], [001], [00-1], directions, at the [100] zone axis. (ii) At the [010] zone axis, the [011] and [0-11] (and [0-1-1]/[01-1]) apices become collinear, producing the straight morphology (symbolized as "s" in (i)). (iii) When tilted, the zigzagged [011] direction appears at the side (symbolized as "z" in (i)), when projected at the

[011] zone axis. (iv) When tilted up to 90°, the [0-11]/[0-1-1] ([011]/[01-1]) apices becomes collinear when projected at the [001] zone axis, producing the straight morphology (symbolized as "s" in (i)).

CONCLUSIONS

In summary, tomographic reconstruction and HRTEM images were used to characterize the unique 3D structure of twinned superlattice Zn$_3$P$_2$ NWs and InAs NWs, synthesized by the vapor transport method. Zn$_3$P$_2$ belongs to a unique tetragonal system, in which its lattice constant (c/a =1.414) leads to its having pseudo cubic symmetry. The Zn$_3$P$_2$ NWs consequently adopt a unique superlattice structure that consists of twinned octahedral slice segments having alternating orientations along the axial [111] direction of the pseudo cubic unit cell. Each octahedral slice segment has its apices indexed as the six equivalent <112> directions at the [111] zone axis. After each 30° turn, the straight and zigzagged morphologies appear repeatedly at the <112> and <011> zone axes, respectively. The 3D structure of the twinned Zn$_3$P$_2$ NWs is virtually the same as that of the twinned ZB InAs NWs. The zigzagged CdO NWs consist of hexahedral segments, whose six apices are matched to the <011> directions linked along the [111] axial direction. The rutile (tetragonal) TiO_2 NBs, grown along the [010] direction, exhibit a unique twisted form that appears at the [011] zone axis, while the wide- and narrow width straight morphologies appear at the [100] and [001] zone axes, respectively. We suggested a twinned superlattice structure that consist of twinned octahedral slices, whose six apices are indexed by the <011>/<001> directions, stacked along the axial [100] direction. The method of tomographic 3D reconstruction, combined with the high resolution TEM images, enables the precise structural analysis of various twinned and zigzagged 1D nanostructures.

ACKNOWLEDGMENT

This research was supported by the WCU (World Class University) program through the NRF funded by the Ministry of Education, Science and Technology (R31-10035).

REFERENCES AND NOTES

1. Hu, J.; Odom, T. W.; Lieber, C. M. *Acc. Chem. Res.* **1999**, *32*, 435.
2. Johansson, J.; Karlsson, L. S.; Svensson, C. P. T.; Mårtensson, T.; Wacaser, B. A.; Deppert, K.; Samuelson, L.; Seifert, W. *Nature Mater.* **2006**, *5*, 574.
3. Verheijen, M. A.; Immink, G.; de Smet, T.; Borgström, M. T.; Bakkers, E. P. A. M. *J. Am. Chem. Soc.* **2006**, *128*, 1353.
4. Tian, M.; Wang, J.; Kurtz, J.; Mallouk, T. E.; Chan, M. H. W. *Nano Lett.* **2003**, *3*, 919.
5. Li, Q.; Gong, X.; Wang, C.; Wang, J.; Ip, K.; Hark, S. *Adv. Mater.* **2004**, *16*, 1436.
6. Davidson, F. M., III; Wiacek, R.; Korgel, B. A. *Chem. Mater.* **2005**, *17*, 230.
7. Chen, H.; Wang, J.; Yu, H.; Yang, H.; Xie, S.; Li, J. *J. Phys. Chem. B* **2005**, *109*, 2573.
8. Ross, F. M.; Tersoff, J.; Reuter, M. C. *Phys. Rev. Lett.* **2005**, *95*, 146104.
9. Ding, Y.; Wang, Z. L.; Sun, T.; Qiu, J. *Appl. Phys. Lett.* **2007**, *90*, 153510.
10. Xiong, Q.; Wang, J.; Eklund, P. C. *Nano Lett.* **2006**, *6*, 2736.
11. Han, W. –Q.; Wu, L.; Stein, A.; Zhu, Y.; Misewich, J.; Warren, J. *Angew. Chem. Int. Ed.* **2006**, *45*, 6554.
12. Shen, G.; Chen, P. –C.; Bando, Y.; Golberg, D.; Zhou, C. *J. Phys. Chem. C.* **2008**, *112*, 16405.
13. Wang, D. -H.; Xu, D.; Wang, Q.; Hao, Y. –J.; Jin, G. –Q.; Guo, X. –Y.; Tu, K. N. *Nanotech.* **2008**, *19*, 215602.
14. Tao, X.; Li, X. *Nano Lett.* **2008**, *8*, 505.
15. Bao, J.; Bell, D. C.; Capasso, F.; Wagner, J. B.; Mårtensson, T.; Trägårdh, J.; Samuelson, L. *Nano Lett.* **2008**, *8*, 836.
16. Meng, Q.; Jiang, C.; Mao, S. X. *J. Cryst. Growth* **2008**, *310*, 4481.
17. Fu, X.; Jiang, J.; Zhang, W.; Yuan, J. *Appl.Phys. Lett.* **2008**, *93*, 043101.
18. Yin, L. W.; Lee, S. T. *Nano Lett.* **2009**, *9*, 957.
19. Wang, Z. W.; Li, Z. Y. *Nano Lett.* **2009**, *9*, 1467.
20. Yang, Y.; Scholz, R.; Fan, H. J.; Hesse, D.; Gösele, U.; Zacharias, M. *ACS Nano* **2009**, *3*, 555.
21. Shen, X. S.; Wang, G. Z.; Hong, X.; Xie, X.; Zhu, W.; Li, D. P. *J. Am. Chem. Soc.* **2009**, *131*, 10812.

22. Dayeh, S. A.; Susac, D.; Kavanagh, K. L.; Yu, E. T.; Wang, D. *Adv. Funct. Mater.* **2009,** *19,* 2102.
23. Ikonić, Z.; Srivastava, G. P.; Inkson, J. C. *Phys. Rev. B* **1993,** *48,* 17181.
24. Verheijen, M. A.; Algra, R. E.; Borgström, M. T.; Immink, G.; Sourty, E.; van Enckevort, W. J. P.; Vlieg, E.; Bakkers, E. P. A. M. *Nano Lett.* **2007,** *7,* 3051.
25. Han, Y.; Zhao, L.; Ying, J. Y. *Adv. Mater.* **2007,** *19,* 2454.
26. Kim, H. S.; Hwang, S. O.; Myung, Y.; Park, J.; Bae, S. Y.; Ahn, J. P. *Nano Lett.* **2008,** *8,* 551.
27. Arslan, I.; Talin, A. A.; Wang, G. T. *J. Phys. Chem. C.* **2008,** *112,* 11093.
28. Ersen, O.; Bégin, S.; Houlle, M.; Amadou, J.; Janowska, I.; Grenèche, J. -M; Crucifix, C.; Pham-Huu, C. *Nano Lett.* **2008,** *8,* 1033.
29. Heigoldt, M.; Arbiol, J.; Spirkoska, D.; Rebled, J. M.; Conesa-Boj, S.; Abstreiter, G.; Peiró, F.; Morante, J. R.; Fontcuberta i Morral, A. *J. Mater. Chem.* **2009,** *19,* 840.
30. Arslan, I.; Tong, J. R.; Midgley, P. A. *Ultramicroscopy* **2006,** 106, 994.
31. Żdanowicz, W.; Kloc, K.; Kalińska, A.; Cisowska, E.; Burian, A. *J. Cryst. Growth* **1975,** *31,* 56.
32. Zanin, I. E.; Aleinikova, K. B.; Afanasiev, M. M.; Antipin, M. Y. *J. Struct. Chem.* **2004,** *45,* 844.
33. Shen, G.; Bando, Y.; Ye, C.; Yuan, X.; Sekiguchi, T.; Golberg, D. *Angew. Chem. Int. Ed.* **2006,** *45,* 7568.
34. Yang, R.; Chueh, Y. -L.; Morber, J. R.; Snyder, R.; Chou, L. -J.; Wang, Z. L. *Nano Lett.* **2007,** *7,* 269.
35. Liu, C.; Dai, L.; You, L. P.; Xu, W. J.; Ma, R. M.; Yang, W. Q.; Zhang, Y. F.; Qin, G. G. *J. Mater. Chem.* **2008,** *18,* 3912.

Solid Oxide Fuel Cells
and Electronic Structure of Ceramics

Mater. Res. Soc. Symp. Proc. Vol. 1262 © 2010 Materials Research Society 1262-W01-02

Phonon density of states of model ferroelectrics

Narayani Choudhury[1, 5], Alexander I. Kolesnikov[2], Helmut Schober[3,6], Eric J. Walter[4], Mark Johnson[3], Douglas L. Abernathy[2], Matthew S. Lucas[2]

[1]Dept. of Physics, University of Arkansas, Fayetteville, AR 72701, USA.
[2]Neutron Sciences Division, Oak Ridge National Laboratory, Oak Ridge, TN 37831, USA.
[3]Institut Laue-Langevin, 38042 Grenoble, Cedex 9, France.
[4]Dept. of Physics, College of William and Mary, Williamsburg, VA 23185, USA.
[5]Solid State Physics Division, Bhabha Atomic Research Centre, Mumbai 400085, India.
[6]Université Joseph Fourier, UFR de Physique, 38041 Grenoble, Cedex 9, France

ABSTRACT

First principles density functional calculations and inelastic neutron scattering measurements have been used to study the variations of the phonon density of states of $PbTiO_3$ and $SrTiO_3$ as a function of temperature. The phonon spectra of the quantum paraelectric $SrTiO_3$ is found to be fundamentally distinct from those of ferroelectric $PbTiO_3$ and $BaTiO_3$. $SrTiO_3$ has a large 70-90 meV phonon band-gap in both the low temperature antiferrodistortive tetragonal phase and in the high temperature cubic phase. Key bonding changes in these perovskites lead to spectacular differences in their observed phonon density of states.

INTRODUCTION

Ferroelectric materials interconvert electrical and mechanical energies and find important applications as piezoelectric transducers and actuators, pyroelectric arrays, non-volatile memories, dielectrics for microelectronics and wireless communication, non-linear optical applications, medical imaging, *etc.* Oxide perovskites like $PbTiO_3$, $BaTiO_3$ and $SrTiO_3$ are well studied [1-10] model ferroelectrics, due to their fundamental interest as well as due to their technological relevance. High voltage photovoltaic effects in ABO_3 perovskites have also been reported [4,5] which suggest possible applications of these oxides as solar energy devices.

$SrTiO_3$ is an incipient ferroelectric with a very large static dielectric response which exhibits unusual phonon anomalies and anomalous electrostrictive response. The phonon instabilities in these perovskites have an important bearing on their piezoelectric and dielectric properties. While the ferrodistortive modes involve zone center phonon instabilities, the antiferrodistortive phases engage zone boundary modes involving rotations of the TiO_6 octahedra [10]. Accurate characterization of the phonon modes in the entire Brillouin zone of these perovskites is therefore highly desirable. These perovskites have a rich phase diagram and their material properties are found to be strongly influenced by their crystal structure and bonding characteristics [1,2]. Both $PbTiO_3$ and $BaTiO_3$ have a simple cubic high temperature paraelectric phase which transforms to a ferroelectric tetragonal phase around 763 K and 403 K, respectively. Tetragonal $PbTiO_3$ is a large strain material (c/a=1.06) exhibiting ferroelectricity up to high temperatures, and it has a single cubic to tetragonal transition. The spontaneous polarization of $PbTiO_3$ at room temperature is almost three times that of $BaTiO_3$. $BaTiO_3$ on the other hand has a much smaller strain (1.01) and exhibits successive ferroelectric phase transitions from paraelectric cubic to tetragonal, orthorhombic and rhombhohedral structures with decreasing temperature. $SrTiO_3$ is a quantum paraelectric which undergoes a transition from the cubic ($Pm\overline{3}m$) to a tetragonal ($I4/mcm$) antiferrodistortive (AFD) phase at 105 K. Experimental studies suggest that although $SrTiO_3$ is paraelectric at low temperatures, it is very close to the ferroelectric threshold with quantum fluctuations of the atomic positions suppressing the ferroelectric instability and leading to a stabilized paraelectric state.

Of particular interest is the complete understanding of the phonon density of states (PDOS). The PDOS is the key quantity required to derive various thermodynamic properties and can be directly measured using inelastic neutron scattering (INS) techniques. Unlike Raman and infrared studies, which probe only the long wavelength phonon excitations, inelastic neutron scattering measurements directly probe the phonon modes in the complete Brillouin zone.

Low temperature (T=6 K) INS measurements and *ab initio* T=0 K density functional theory (DFT) studies reveal that the PDOS of the quantum paraelectric $SrTiO_3$ is fundamentally distinct from those of ferroelectric $PbTiO_3$ and $BaTiO_3$ [1] with a large 70-90 meV phonon band gap. Here, we report *in situ* studies of the changes in the phonon density of states of ferroelectric $PbTiO_3$ and the quantum paraelectric $SrTiO_3$ with temperature. The data are interpreted using *ab initio* first principles calculations.

METHODS

First principles density functional theory (DFT) and density functional perturbation theory (DFPT) linear response calculations [1] have been carried out using plane wave basis sets and the code ABINIT [11] within the local density approximation (LDA) for electron exchange and correlation. These studies employ norm conserving pseudopotentials obtained using the code OPIUM [12]. The Brillouin zone integrations were carried out with a 6x6x6 k-point mesh using a plane wave energy cut off of 50 Ha. Exact linear response calculations on a 4x4x4 q-point grid of wavevectors were undertaken and the force constants derived. Additional details are as given in Ref. [1].

The inelastic neutron scattering measurements were carried out on the High Resolution Medium Energy Chopper Spectrometer (HRMECS) at the Intense Pulsed Neutron Source of the Argonne National Laboratory. The energy resolution ΔE of the HRMECS spectrometer varies between 2–4% of the incident-neutron energy over the neutron-energy-loss region. Two incident neutron energies were used (E_i=50 and 130 meV) to get a good resolution in all range of energy transfers. The data were collected in wide range of scattering angles (28 to 132 degrees), which provides large coverage of momentum transfer to insure good averaging of the vibrational spectra over inverse space. The data were corrected for background scattering by subtracting the results from empty container runs. Measurements of the incoherent scattering from a vanadium standard provided the detector calibration and intensity renormalization. We used commercially purchased samples (Sigma-Aldrich, > 99 % purity) for these studies. The samples were first cooled to 6 K by a closed cycled refrigerator with the sample temperature maintained within 0.1 K throughout a run to minimize multiphonon contributions. The samples were then heated in a furnace to obtain desired temperatures. Detailed analysis of the inelastic-neutron-scattering data in the incoherent approximation [1] were undertaken for comparison with the calculated spectra.

RESULTS AND DISCUSSION

The theoretical *ab initio* studies reveal that at ambient pressure and T=0 K, tetragonal $PbTiO_3$ (space group *P4mm*) is dynamically stable in the entire Brillouin zone. On the other hand, zero pressure LDA calculations of paraelectric cubic $SrTiO_3$ yields soft zone center and zone boundary R-point phonon mode instabilities in agreement with earlier studies [10]. These soft mode instabilities exhibit strong volume dependence. The phonon density of states which involves integration over the phonon modes in the entire Brillouin zone is however not significantly altered by the changes in volume which significantly affects the dynamical stability of these modes. LDA calculations on a coarse wavevector mesh of the tetragonal AFD phase suggest that the phonon density of states of the tetragonal (AFD) and cubic paraelectric phases are overall quite similar.

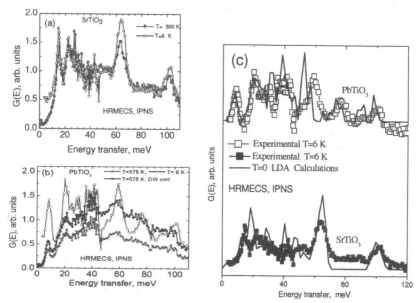

Figure 1. The measured inelastic neutron spectra of the quantum paraelectric $SrTiO_3$ (a) and ferroelectric $PbTiO_3$ (b) as a function of temperature. (c) The T=6 K measured INS data are compared with T=0 *ab initio* density functional calculations and found to be in good agreement. $SrTiO_3$ has an AFD tetragonal structure at low temperature and is cubic above 104 K. The Curie temperature of $PbTiO_3$ is around 760 K. The INS spectra shown in (b) for $PbTiO_3$ are in the ferroelectric phase and the data corrected for Debye Waller factors at T=575 K are also shown for comparison. Additional corrections of the high temperature neutron data of $PbTiO_3$ (b) to incorporate multiphonon effects are currently in progress. The vibrational spectra of the quantum paraelectric $SrTiO_3$ is fundamentally distinct from that of ferroelectric $PbTiO_3$, with a 70-90 meV phonon band gap [1]. These differences in phonon spectra are due to key changes in bonding in the ferroelectric and paraelectric phases [1].

Inelastic neutron scattering measurements at 6 K and 300 K (Fig. 1(a)) carried out at the Intense Pulsed Neutron Source Facility indeed confirm these theoretical predictions and the spectra in the cubic and antiferrodistortive phases are found to be very similar (Fig. 1(a)). The tetragonal AFD to cubic transition of $SrTiO_3$ has a non-polar character and does not affect its dielectric properties.

The computed neutron-weighted generalized phonon density of states of tetragonal $PbTiO_3$, and cubic and tetragonal $SrTiO_3$ are found to be in excellent agreement with the measured inelastic neutron scattering data (Fig. 1). The Pb and Sr vibrations span the low energy 0-40 meV range, while the Ti and O vibrations span the 40-80 and 0-120 meV, respectively [1]. The changes in the phonon spectra in the low energy region involve mainly the Pb and Sr vibrations. In Fig. 2, we display our computed charge densities in these perovskites. Electronic structure calculations [2] revealed that the hybridization between the titanium $3d$ states and the oxygen $2p$ states is essential for ferroelectricity in $PbTiO_3$ and $BaTiO_3$. While the Pb-O bond is covalent in $PbTiO_3$, the Ba-O and Sr-O bonds of $BaTiO_3$ and $SrTiO_3$ are ionic. The computed charge density distributions are in good agreement with the observed distributions of tetragonal $PbTiO_3$ and $BaTiO_3$ obtained using maximum entropy analysis of synchrotron data [13]. The strong covalency of the Pb-O bonds in tetragonal $PbTiO_3$ which arises from the hybridization of the Pb $6s$ state and O $2p$ state has been theoretically predicted [2] as a key factor of the much larger ferroelectricity of $PbTiO_3$ as compared to $BaTiO_3$. The bonding changes in these materials (Fig. 2) also manifest in their dynamical properties [1] and lead to important differences in the phonon density of states (Fig. 1). Whereas, the quantum paralectric $SrTiO_3$ has a large 70-90 meV phonon band gap in the AFD tetragonal and cubic phases, this gap is bridged by the covalent Ti-O vibrations found essential for ferroelectricity. The strong covalent bonding of $PbTiO_3$ gives rise to important anisotropies in the computed and observed vibrational, elastic, piezoelectric and dielectric properties [1].

With increasing temperature there are radical changes in the phonon spectra of ferroelectric $PbTiO_3$ (Fig. 1). The high temperature data of $PbTiO_3$ are found to be in good agreement with extensive studies carried out using the IN6 spectrometer at the Institut Laue Langevin and the ARCS spectrometer at the Spallation Neutron Source, Oak Ridge National Laboratory. These studies reveal that indeed the phonon spectra in the paraelectric phase of $PbTiO_3$ has a band gap around the 70 meV range as obtained for $SrTiO_3$. These details will be reported elsewhere.

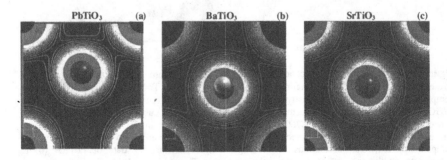

Figure 2. The computed first principles derived charge density distributions (displayed using the software xcrysden [14]) of ferroelectric tetragonal $PbTiO_3$ and rhombhohedral $BaTiO_3$ compared with that of paraelectric cubic $SrTiO_3$. The charge densities were computed from the fully relaxed zero pressure LDA structures. Various details are as given in Ref. [1].

CONCLUSIONS

A combination of first principles calculations and inelastic neutron scattering experiments have been helpful in understanding the temperature variations of the phonon spectra of $PbTiO_3$ and $SrTiO_3$. Our studies provide insights into how changes in bonding in $PbTiO_3$ and $SrTiO_3$ which radically alters their material properties manifest in their phonon density of states. The spectacular differences of the phonon spectra obtained in the ferroelectric and paraelectric phases have important implications for use of vibrational spectroscopy for novel materials design [1].

ACKNOWLEDGEMENTS

These studies were carried out using the supercomputing resources of the Bhabha Atomic Research Centre (N.C.) and the Center for Piezelectrics by Design, College of William and Mary (E.J.W.). Narayani Choudhury acknowledges financial support from the ILL and NMI3 (European Union, Project Number FP7-INFRASTRUCTURES-2008-1 (226507)) for travel to the ILL for participating in these experiments and for attending the MRS Spring meeting at San Francisco. Work at ORNL was supported by the DOE-BES and was managed by UT-Battelle, LLC, for DOE under Contract DE-AC05-00OR22725, and work at ANL was performed under the auspices of the DOE-BES under contract DE-AC02-06CH11357.

REFERENCES

1. N. Choudhury, E.J. Walter, A.I. Kolesnikov, and C.K. Loong, Phys. Rev. B77, 134111 (2008) and references therein.
2. R.E. Cohen, Nature (London) 358, 136 (1992).
3. D. Vanderbilt, Current Opinion in Solid State and Materials Science 2, 701 (1997), and references therein.
4. K. Uchino, Y. Miyazawa and S. Nomura, Jpn. J. Appl. Phys. 21, 1671(1982).
5. J. W. Bennett, I. Grinberg and A. M. Rappe, J. Am. Chem. Soc. 130, 17409 (2008) and references therein.
6. P. Ghosez, E. Cockayne, U. V. Waghmare, and K. M. Rabe, Phys. Rev. B60, 836 (1999).
7. J.D. Freire and R.S. Katiyar, Phys. Rev. B37, 204 (1988), and references therein.
8. C.M. Forster, M. Grimsditch, Z. Li and V. G. Karpov, Phys. Rev. Lett. 71, 1258 (1993).
9. G. Shirane, Rev. Mod. Phys. 46, 437 (1974) and references therein.
10. N. Sai and D.Vanderbilt, Phys. Rev. B62, 13943 (2000).
11. X. Gonze et al., Comp. Mat. Sci. 25, 478 (2002); http://www.abinit.org/
12. http://www.sourceforge.net/
13. Y. Kuroiwa, S Aoyagi, A. Sawada, J Harada, E. Nishibori, M. Takata, M. Sakata Phys. Rev. Lett. 87, 217601 (2001).
14. A. Kokalj, Comp. Mater. Sci. 28, 155 (2003).

Solid Oxide Fuel Cells and Solid State Ionics

Mater. Res. Soc. Symp. Proc. Vol. 1262 © 2010 Materials Research Society   1262-W02-06

NEUTRON DIFFRACTION STUDY OF NiO/LiCoO₂ ELECTRODES FOR INNOVATIVE FUEL CELL DEVELOPMENT

R. Coppola [1], P. F. Henry [2][4], A. Moreno [3], J. Rodriguez-Carvajal [2], E. Simonetti [3]

[1] ENEA-Casaccia, UTFISSM, Via Anguillarese 301, 00123 Roma, Italy

[2] ILL, 6, rue Jules Horowitz, 38042 Grenoble, France

[3] ENEA-Casaccia, UTRINN, Via Anguillarese 301, 00123 Roma, Italy

[4] Helmholtz Zentrum Berlin, Hahn-Meitner-Platz 1, 14109, Berlin, Germany

Abstract

Ni-NiO 30 wt% electrodes coated with $LiMg_{0.05}Co_{0.95}O_2$ cobaltite deposited on the substrate by complex sol-gel process were investigated using powder neutron diffraction. As the catalytic layer is only 1-2 microns thick, the diffracting volume of the cobaltite phase was increased by stacking 20 small rectangular pieces cut from the original electrode, assuming that the catalytic layer on the electrodes was homogenous. A pure bulk cobaltite sample was used as a reference for identifying the diffraction peaks of the catalytic layer in the complete electrode. Both an as-received sample and a tested electrode (100 h at 650 °C in a cell) were measured. Despite the small diffracting volume of the catalytic layer, due to the high flux available at D20, it was possible to detect the hexagonal phase and estimate its volume fraction in the as-received sample. In the tested electrode, the cobaltite material is no longer present, while traces of magnesium oxide and cobalt oxide are detected, suggesting significant modifications take place during use.

1. Introduction

Due mainly to its fast oxygen reduction reaction rate $Li_xNi_{1-x}O$ is considered as one of the best suited materials for the development of innovative Molten Carbonate Fuel Cells. However, due to its high rate of dissolution in molten carbonate, the cathodic nickel is dispersed and transported inside the electrolyte producing a concentration gradient under the electric field of the cell and a considerable decrease of the estimated life time. In order to avoid or to limit this effect, the electrode can be coated by a thin layer of material characterized by a lower solubility in molten carbonates such as ferrites, manganites and cobaltites (1-2) and more specifically lithium cobaltite ($LiCoO_2$), which is commercially employed in lithium-ion batteries and in electrochromic films (3, 4). A process for the production of Mg doped $LiCoO_2$ powders and the preparation of porous cathodes of 100 cm^2 in size has been developed (3-9). The porous nickel cathode has been coated by a thin layer of Mg-doped $LiCoO_2$; the complex sol-gel process technique was selected because of its effectiveness in covering the porous substrate deeply in the micro- and macro-holes. The morphology, crystallographic structure, conductivity, solubility of such electrodes have been investigated; in-cell tests have been carried out obtaining encouraging results from the electrochemical point of view in comparison with nickel oxide (9). Within this scientific frame, neutron diffraction has been utilized to characterize the crystallographic structure of the metallic substrate and of the catalytic layer, both in the as-received state and after exposure to significant service conditions. In fact, neutron diffraction provides a powerful tool for such investigations (10) and has also been utilized to investigate solid state reaction synthesized $LiCoO_2$ (11), similar to the one considered in this work.

2. Material characterization

The cathodes were prepared by means of a sol-gel process first adding LiOH to aqueous acetates solution of Co $^{2+}$ with ascorbic acid, then alkalizing with aqueous ammonia to pH=8 (12). In these sols, diluted with ethanol, porous cathode plates were dipped and withdrawn at controlled rate several times to achieve the required film thickness. The coated substrate was soaked at 200 °C for 72 h, then at 400 °C for 1 h and calcined (using a low heating and cooling rate of 1°C min^{-1}) at 650 °C for up to 4 h. During the heat treatments Ni plates were always placed between ceramic sheets to avoid warping. The thermal treatment procedure was determined on the basis of thermal analysis data. The formation of the $LiCoO_2$ crystalline phase on the porous nickel surface was checked by X-ray diffraction and scanning electron microscopy (9). In-cell tests showed a good cell performance that gradually improved during the cell operation time, with a voltage of 800 mV at a current density of 100 mA cm^{-2} after 700 h. The diffracting volume of the catalytic layer is only a few microns thick on each side of the Ni/NiO plate, which is itself approximately 0.5 mm thick. Therefore, the original 10×10 cm^2 electrode was cut into 20 small rectangular pieces (1×5 cm strips), which were than stacked together obtaining a sufficient estimated $LiCoO_2$ mass in the volume investigated by the neutron beam. The composition of the electrode substrate was Ni/NiO 30 wt%, while the catalytic layer was $LiCo_{0.95}Mg_{0.05}O_2$. Both an as-received electrode and an identical one tested at 650°C for 100 h were investigated. Reference samples were also prepared to separately characterize the crystallographic structure of the electrode (pure Ni and Ni/NiO 30 wt%) and of the catalytic layer, for which tape cast $LiCo_{0.95}Mg_{0.05}O_2$ cobaltite was taken.

3. Experimental technique and results

The neutron diffraction measurements were carried out at the D20 diffractometer at the High Flux Reactor of the Institut Max von Laue – Paul Langevin (13). The incident wavelength of the instrument was varied and an optimum wavelength of 1.3 Å chosen based on d-spacing range and flux. In the cases where the samples were in the form of strips, two different data collection geometries were used. The first one involved mounting the strips in bundles with the individual strips parallel to the incoming neutron beam and the second with the strips aligned perpendicular to the beam. This allowed the effects of preferred orientation of the deposited material to be probed. The differences observed in the patterns obtained as a result of preferred orientation do not substantially affect the refined Ni/NiO and cobaltite models. The data were refined using Fullprof (14) and GSAS (15).

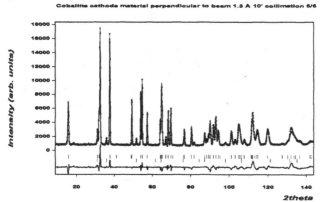

Figure 1. Final Rietveld profile pattern of the tape cast pure cobaltite cathode material where the circles represent the data, the continuous line the final fit, the lower continuous line the difference plot and the tick marks the allowed reflections. The upper tick marks are $LiMg_xCo_{1-x}O_2$ (x=0.05) with a volume fraction of 98.04(13)% and the lower residual MgO.

Figure 1 shows the diffraction pattern of the tape-cast cobaltite cathode material $LiCo_{0.95}Mg_{0.05}O_2$: there is an impurity phase present in the pattern and a search of the possible impurities showed that MgO is a good fit for residual non-fitted peaks. The MgO volume fraction was refined to be 1.96(13)%, compared with the value of 0.65 wt% found in a previous study for the solid state reaction synthesized material with identical nominal composition (11). There are also two low angle peaks that are not fitted to either phase but could not be assigned to any known oxides. The atomic positions are given in Table I and are in good agreement with those found in the previous work (11).

Table I Refined atomic parameters from the bulk cobaltite cathode $LiMg_xCo_{1-x}O_2$ material mounted perpendicular to the incident neutron beam in the space group R -3 m (hexagonal setting) with lattice parameters $a = 2.84005(25)$ Å; $c = 14.1783(8)$ Å; $\gamma = 120\,°$.

Atom	X	Y	Z	Frac	U_{iso} x 100/ Å²
Li	0	0	0.5	1.0	0.71
Mg	0	0	0	0.05*	0.06
Co	0	0	0	0.95*	0.06
O	0	0	0.26017(6)	1.0	0.25

* fixed composition

Figure 2 gives the final Rietveld fit to the diffraction pattern of the as-received deposited electrode, showing extra reflections that cannot be attributed to the original electrode material, that correspond well with those for $LiMg_xCo_{1-x}O_2$ (x=0.05). The powder neutron diffraction pattern and partial Rietveld fit of the treated electrode are shown in Figure 3. NiO is clearly present but the original cobaltite material is not, replaced by peaks that do not correspond to phases known to be present in bulk samples. The technique is however sensitive enough to detect traces of magnesium oxide and cobalt oxide.

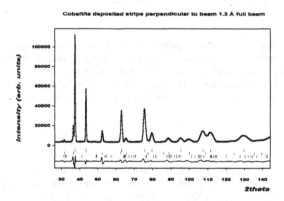

Cobaltite deposited strips perpendicular to beam 1.3 A full beam

Figure 2. Final Rietveld profile pattern of the as-received deposited cathode material where the circles represent the data, the continuous line the final fit, the lower line the difference plot and the tick marks the allowed reflections. The upper tick marks are Ni, middle NiO and bottom $LiMg_xCo_{1-x}O_2$ with volume fractions of 81.7(10), 17.4(5) and 0.9(6)%

38

respectively.

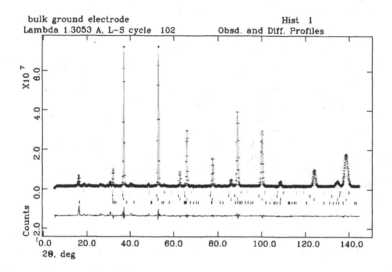

Figure 3. Rietveld profile pattern fit of the treated electrode (crosses) where the marked phases are as follows from top to bottom: $LiMg_xCo_{1-x}O_2$, nickel oxide, magnesium oxide and cobalt oxide. The only well visible phase present is nickel oxide.

4. Discussion/Conclusions

Ni/NiO electrodes coated by a few micron thick layer of sol-gel deposited $LiCo_{0.95}Mg_{0.05}O_2$ have been investigated using powder neutron diffraction. In the as-received electrode, a cobaltite volume fraction of the order of 1% was estimated. After 100 h at 650°C in the cell, the catalytic layer is completely changed and only traces of hexagonal phase are detectable. The technique is of use as it allows bulk probing of the deposited electrode as a whole in a non-destructive way. However, the origin of the changes could not be determined from such an ex-situ type measurement but show that exploitable data can be obtained from the cathode materials in short timescales using neutron diffraction instrumentation. In order to characterise the electrodes fully and the processes leading to the observed phase changes, an in-situ study using a full cell set up is required. Complementary X-ray diffraction and EXAFS measurements are also required to provide information on the surface distribution of the phases present in the tested electrode and on the homogeneity.

References

1. Y. Miyazaki, M. Yanagida, S. Tanase, K. Tanimoto, T. Kojima, N. Ohtori, H.Okuyama, T. Kodama, K. Morimoto, I. Nagashima, C. Nagai, H. Itoh, *Proc. 1992 Fuel Cell Seminar*, Nov. 29 - Dec. 2, 1992, Tucson (USA)
2. Y. Miyazaki, M. Yanagida, K. Tanimoto, T. Kojima, N. Ohtori, T. Asai, *Proc. 1994 Fuel Cell Seminar*, Nov. 28 - Dec. 1, 1994, S. Diego (USA)
3. T. Fukui, H. Okawa, K. Kodera, T. Tsunooka, *Proc. 2nd International Fuel Cell Conference*, Kobe, February 1996
4. A. Lundblad, *Doctoral Thesis*, University of Stockholm, 1996
5. M. Carewska, F. Pallini, S. Scaccia, ENEA Report RTI/ERG/TEA/96(02)
6. L. Giorgi, M. Carewska, M. Patriarca, S. Scaccia, E. Simonetti and A. Di Bartolomeo, *Journal of Power Sources*, 49 (1994) 227 - 243
7. L. Giorgi, M. Carewska, S. Scaccia, E. Simonetti, F. Zarzana, *Denki Kagaku*, 6 (1996) 482
8. L. Giorgi, M. Carewska, E. Simonetti, S. Scaccia, F. Croce, A. Pozio, in *Electrochemical Technology of Molten Salts*, eds. C.A.C. Sequeira, G.S. Picard, pp. 285-302, *Trans Tech Publications*, Switzerland, 1993
9. E. Simonetti, R. Lo Presti, *Journal of Power Sources*, 160 (2006) 816
10. Y. Chabre, J. Pannetier, *Progr. Sol. St. Chem*, 23 1 (1995) 1
11. R. Coppola, F. Bourée, L. Giorgi, *Physica B*, 276-278(2000) 862
12. A. Deptula, W. Lada, *J. New. Mat. Electrochem. Systems* 6,33-37 (2003)
13. T.C. Hansen, P.F. Henry, H.E. Fischer, J. Torregrossa, P. Convert. *Meas. Sci. Technol.* 2008, **19(3)**, 034001.
14. J. Rodríguez-Carvajal, Recent Developments of the Program FULLPROF, in Commission on Powder Diffraction (IUCr).Newsletter (2001), 26, 12-19.
15. Larson, A. C. & Von Dreele, R. B. (1994). General Structure Analysis System (GSAS). Report LAUR 86-748. Los Alamos National Laboratory, New Mexico, USA

H2 Storage and Hydrogen in Solids I

Mater. Res. Soc. Symp. Proc. Vol. 1262 © 2010 Materials Research Society   1262-W03-03

Quaternary Ammonium Borohydride Adsorption in Mesoporous Silicate MCM-48

Michael J. Wolverton[1,2], Luke L. Daemen[1], Monika A. Hartl[1]
1. Los Alamos Neutron Science Center, Los Alamos National Laboratory,
Los Alamos, NM 87545, USA
2. Dept. of Applied Science, University of Arkansas at Little Rock, Little Rock, AR 72204, USA

ABSTRACT

Inorganic borohydrides have a high gravimetric hydrogen density but release H_2 only under energetically unfavorable conditions. Surface chemistry may help in lowering thermodynamic barriers, but inclusion of inorganic borohydrides in porous silica materials has proved hitherto difficult or impossible. We show that borohydrides with a large organic cation are readily adsorbed inside mesoporous silicates, particularly after surface treatment. Thermal analysis reveals that the decomposition thermodynamics of tetraalkylammonium borohydrides are substantially affected by inclusion in MCM-48. Inelastic neutron scattering (INS) data show that the compounds adsorb on the silica surface. Evidence of pore loading is supplemented by DSC/TGA, XRD, FTIR, and BET isotherm measurements. Mass spectrometry shows significant hydrogen release at lower temperature from adsorbed borohydrides in comparison with the bulk borohydrides. INS data from partially decomposed samples indicates that the decomposition of the cation and anion is likely simultaneous. These data confirm the formation of Si-H bonds on the silica surface upon decomposition of adsorbed tetramethylammonium borohydride.

INTRODUCTION

Large amounts of hydrogen can be released through the thermal decomposition of hydride complexes [1,2]. As the projected hydrogen density requirements for hydrogen storage media have become more demanding, borohydrides have gained popularity as a potential storage medium [3]. The thermodynamic parameters obtained via *ab-initio* studies predict that many borohydrides are theoretically capable of thermolysis at very reasonable temperatures [4,5]. In practice, much higher temperatures are required, and the rate of reaction is often extremely sluggish except in the presence of a catalyst [1,2,6]. This implies that the limiting factors are associated with kinetics and the activation energy barrier. Silica has been shown to encourage the decomposition of $LiBH_4$ causing the reaction to occur at a lower temperature, with a faster rate of gas release [1]. Fang et. al. have shown that that dispersing $LiBH_4$ with a carbonaceous mesoporous network encourages thermal decomposition [7]. The degree of inclusion within the pores was somewhat ambiguous in the Fang et. al. study. Conclusive evidence of borohydride encapsulation in a mesoporous silicate has never been reported in the literature. MCM-48 was chosen for this study because of the regularity of its pore size, high specific surface area, and cubic pore structure.

Tetraalkylammonium borohydrides (R_4NBH_4) were selected for MCM-48 loading because common alkali metal borohydrides such as $NaBH_4$ and $LiBH_4$ do not properly enter the pores from solution. This is discussed in further detail as proper pore loading of R_4NBH_4 is verified.

EXPERIMENT

Incoherent inelastic neutron scattering (IINS) provides the advantage of extreme sensitivity to ^{1}H hydrogen, and penetration of the neutrons through the silica network. The mesoporous silicate contributes little to the vibrational spectrum when a hydrogen rich tetraalkylammonium borohydride is present. INS spectra were collected on the Filter Difference Spectrometer (FDS) at the Manuel Lujan, Jr. Neutron Scattering Center at Los Alamos National Laboratory. Samples were sealed in aluminum sample cans under helium and maintained at 10K during INS analysis to reduce the thermal broadening of vibrational modes. X-ray diffraction (XRD) patterns were collected on a Rigaku Ultima III equipped with a vacuum cuff attachment. XRD scans were performed under vacuum. Mass spectra (MS/RGA) were collected with an SRS RGA-300 residual gas analyzer fitted with a two stage vacuum dosing manifold maintaining a pressure range of 10^{-10} - 10^{-5} torr. Fourier transform infrared (FTIR) spectra were collected with a Thermo-Nicolet Nexus 670 FTIR equipped with an ATR accessory. The determination of surface area was accomplished via the fit of nitrogen adsorption data at 77K with Brunauer-Emmett-Teller (BET) isotherms. These data were recorded on a Micromeritics Gemini V, and BET isotherms were fitted using the accompanying software. Thermogravimetric analysis (TGA) data were recorded on a Netzsch Jupiter STA449C. Samples were maintained under a light flow of Ar during TGA measurements.

MCM-48 was synthesized in pressure vessel by hydrothermal methods as described elsewhere [8]. Before loading the pores with borohydride, the surface of MCM-48 was at least partially functionalized with methyl groups. Methylation is accomplished by stirring the calcined MCM-48 in toluene with trimethylchlorosilane, Me$_3$ClSi, under argon. The MCM-48 is further washed with toluene, and dried under vacuum overnight. Methylation reduces the number of surface silanol groups (Si-OH) and makes the surface less hydrophilic. This provides slightly better resistance to atmospheric moisture contamination when handling samples in air. Experience suggests that loading of the pores with tetraalkylammonium borohydrides occurs slightly more readily with a methylated surface.

Tetramethylammonium borohydride, (CH$_3$)$_4$NBH$_4$, was prepared by ion exchange of tetramethylammonium hydroxide with sodium borohydride in water according to the following reaction.

$$Me_4NOH + NaBH_4 \xrightarrow{H_2O} Me_4NBH_4 + NaOH$$

The pentahydrate of tetramethylammonium hydroxide was obtained from Fischer Scientific and NaBH$_4$ from ACROS. Reagents were used as received. NaOH was removed from the product with ethanol, in which Me$_4$NBH$_4$ is sparsely soluble. FTIR analysis shows the prepared Me$_4$NBH$_4$ to be free of moisture and ethanol.

Tetraethylammonium borohydride, (C$_2$H$_5$)$_4$NBH$_4$, was prepared by ion exchange of tetraethylammonium bromide with sodium borohydride in wet dichloromethane (DCM).

$$Et_4NBr + NaBH_4 \xrightarrow{DCM/H_2O} Et_4NBH_4 + NaBr$$

Tetraethylammonium bromide and sodium borohydride were obtained from ACROS and used as received. Ion exchange was performed by dissolving a stoichiometric mixture of reagents in 10:1 DCM / H$_2$O which was first made slightly alkaline with NaOH. After the initial wet DCM solvent mixture is removed under vacuum, product separation is accomplished by dissolving Et$_4$NBH$_4$ in anhydrous DCM. XRD of the purified Et$_4$NBH$_4$ revealed no presence of the byproduct NaBr, and the final product was verified to be free of DCM by FTIR analysis.

Loading of the MCM-48 is accomplished by the vacuum distillation of a concentrated solution of each tetraalkylammonium borohydride. Solvents were chosen carefully so that they could be removed at relatively low temperature under vacuum. Solvent which strongly adsorb onto MCM-48 tend to cause pore clogging and inhomogeneous loading as the solute crashes out before it enters the network. This is the primary aforementioned difficulty with ether solutions of in alkali borohydrides. Methanol was used for Me_4NBH_4, and DCM for Et_4NBH_4. Prior to loading, it was verified via TGA that full removal of both methanol and DCM from the methylated MCM-48 was possible at ~30°C with a turbomolecular vacuum pump.

DISCUSSION

Prior to INS analysis, it was verified by supporting analyses of XRD, BET and FTIR that the characteristics of the MCM-48 samples loaded with R_4NBH_4 were consistent with the loading of the pores of MCM-48, rather than a mixture. For brevity, some of these supporting data will be shown only for MCM-48 loaded with Et_4NBH_4. However, it should be noted that all identified trends occur with the MCM-48 loaded with Me_4NBH_4 samples as well.

The only notable features in the XRD pattern of MCM-48 are at low scattering angles. A Bragg peak around $2\theta \approx 2.5°$ is caused by the regularly spaced pores in the MCM-48 network as can be seen in Figure 1. This peak is reduced in intensity upon loading of MCM-48 with an organic borohydride. The reduction in intensity is caused by the decrease in scattering contrast between the pore and silica framework when the pore is loaded with a scatterer, namely the borohydride. The peak is not shifted, because the location of the Bragg peak corresponds to the center-to-center spacing of the pores. The reduced peak intensity is reliably reproducible across multiple samples of the same loading batch, indicating that the loading of the pores is relatively uniform and homogeneous.

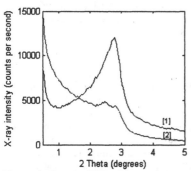

Figure 1. Powder XRD (low angles).
1) Methyl MCM-48. 2) MCM-48 loaded with Et_4NBH_4.

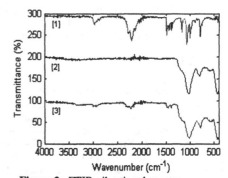

Figure 2. FTIR vibrational spectra.
1) Bulk Et_4NBH_4. 2) Methyl MCM-48.
3) Methyl MCM-48 loaded with Et_4NBH_4.

Supplementary evidence of pore loading is the IR spectrum of Et_4NBH_4 loaded in MCM-48 shown in Figure 2. After loading in MCM-48, the vibrational modes of Et_4NBH_4 are still present, but highly attenuated. However, the SiO_2 modes from the MCM-48 appear at full strength. This would not be the case if the Et_4NBH_4 were merely coating the outside of MCM-48 particles. The attenuation of the Et_4NBH_4 peaks is due to the attenuation of IR radiation through silica.

The BET specific surface area of the methylated MCM-48 material was measured as ~1200 m^2/g before loading, and ~350 m^2/g after loading with Et_4NBH_4. The reduction in specific surface area is consistent with significant pore loading, although clogging of the entrance of the pores cannot be excluded on the basis of this measurement alone.

Figures 3 and 4 show TGA data for pure R_4NBH_4, and R_4NBH_4 loaded in MCM-48 for R=Me and R=Et respectively. It should be noted that methylated MCM-48 (not shown) displays measurable mass loss until temperatures well above 500°C.

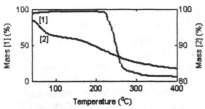 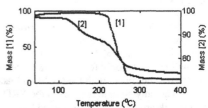

Figure 3. TGA 5K/min. 1) Bulk Me_4NBH_4. 2) Methyl MCM-48 loaded with Me_4NBH_4.

Figure 4. TGA 5K/min. 1) Bulk Et_4NBH_4. 2) Methyl MCM-48 loaded with Et_4NBH_4.

The thermal analyses in Figures 3 and 4 reveal that temperature for the onset of decomposition is lowered significantly for both borohydrides when adsorbed in MCM-48. Mass spectrometry reveals that the primary product of both decomposition reactions is hydrogen. However, the mass spectrum also indicates that a significant quantity of diborane is present in the decomposition gas stream of both R_4NBH_4 compounds. It should be noted that the relative quantities of diborane and hydrogen were nearly identical for the bulk R_4NBH_4 compounds and compounds encapsulated in MCM-48. It would seem that diborane formation was neither suppressed nor enhanced by interaction with the MCM-48 surface.

RESULTS

Encapsulation of Me_4NBH_4 in MCM-48 affects the vibrational spectrum substantially, as can be seen in the INS data shown in Figure 5. The background contributions of FDS and the sample can were accounted for by subtracting the spectrum of a vanadium rod in an identical aluminum sample can from the methyl MCM-48 spectrum. In all spectra measured from MCM-48 loaded with Me_4NBH_4, the methyl MCM-48 spectrum was subtracted as background. Shifts in frequency and changes in intensity of the vibrational modes provide further indication that the Me_4NBH_4 is indeed adsorbed onto the silica surface [9]. The wavenumber values at peaks of corresponding features for bulk and adsorbed Me_4NBH_4 respectively are given the table accompanying Figure 5. Assignments were made based on the INS analysis of Me_4NBH_4 (and isotopic variations) of Eckert et. al. supplemented by calculations of the free $[BH_4]^-$ and $[Me_4N]^+$ ion normal modes with Gaussian03[10]. Geometry optimization and vibrational modes were calculated at the MP2 level of theory using the 3-21g basis set.

In the $[BH_4]^-$ anion, the librational mode that occurs at ~215 cm^{-1} in neat Me_4NBH_4 is shifted slightly to higher frequency, and broadened. This is likely due to increased restoring force on the anion caused by the silica surface potential and is indicative of anion adsorption. In the

[Me$_4$N]$^+$ cation, the scissors bending mode (at ~360 cm^{-1} in neat Me$_4$NBH$_4$) and methyl torsions (at ~265 cm^{-1} and 280 cm^{-1} in neat Me$_4$NBH$_4$) are significantly shifted in frequency and altered in intensity. Neither the umbrella deformation mode (at 460 cm^{-1} in neat Me$_4$NBH$_4$) nor the methyl rocking modes of the cation are affected significantly. Methyl torsions are very sensitive to their molecular environment; higher frequency internal modes much less so. The frequency shifts and intensity changes are further indications of interaction of [Me$_4$N]$^+$ with the pore surface.

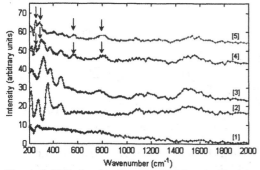

Wavenumber (cm^{-1})	
[2] Bulk Me$_4$NBH$_4$	[3] Adsorbed Me$_4$NBH$_4$
215	223
276	321
353	374
460	460
1079	1079
1104	1104
~1200	~1200
~1400-1600	~1400-1600

Figure 5. INS vibrational spectra. 1) Methyl MCM-48. 2) Bulk Me$_4$NBH$_4$. 3) Methyl MCM-48 loaded with Me$_4$NBH$_4$. 4) Sample 3 after 200°C in vac. 30 min. 5) Sample 4 after 300°C in vac. 30 min.

Upon decomposition, there is a decrease in relative intensities of peaks corresponding to both the cation and anion, indicating that the ions decompose simultaneously. It is possible that the decomposition of the cation and anion are concerted. Immediately visible upon even partial decomposition of adsorbed Me$_4$NBH$_4$, and persisting through the complete decomposition are new spectral features marked with arrows in Figure 5. These peaks also appear in the neutron vibrational spectrum of methylated MCM-48 (or even methylated amorphous silica, 250 m2/g) upon exposure to H$_2$ at 100°C and 350 psi. These features are due to the formation of a surface Si-H bond.

The vibrational spectrum of Et$_4$NBH$_4$, is also affected by adsorption in MCM-48, as shown in the INS spectra displayed in Figure 6. In all spectra measured from MCM-48 loaded with Et$_4$NBH$_4$, the methyl MCM-48 spectrum was subtracted as background. The primary effects of the mesoporous silicate host on the INS spectrum of Et$_4$NBH$_4$ occur below 800 cm^{-1}, with little visible effect on the higher frequency modes. The wavenumber values at peaks of corresponding features for bulk and adsorbed Et$_4$NBH$_4$ respectively are given the table accompanying Figure 6. Gaussian simulations of the [Et$_4$N]$^+$ cation (MP2 level of theory, 3-12g basis) were used to identify the vibrational modes corresponding to the perturbed ranges. Modes in the 230cm^{-1} range are primarily dominated by torsion and rocking of the terminal CH$_3$ groups on each ethyl group. Vibrational motions corresponding to the spectral features in the 380 cm^{-1} range appear to be due primarily to N-C-C deformations. Upon adsorption, a new peak appears at 685 cm^{-1}. This peak persists through the complete decomposition of the Et$_4$NBH$_4$. It is likely that this is a surface compound formed by partial dissociation of the [Et$_4$N]$^+$ cation. The frequency shifts and intensity changes in several vibrational modes of the cation and anion are

47

once again an indication that Et_4NBH_4 adsorbs on the surface, which then plays a role in the thermal decomposition of the material.

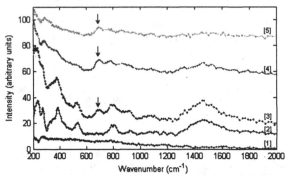

Wavenumber (cm^{-1})	
[2] Bulk Et_4NBH_4	[3] Adsorbed Et_4NBH_4
234	Merge below
270	250
382	374
533	527
No feature	685
793	793
920	920
~1470	~1470

Figure 6. INS vibrational spectra. 1) Methyl MCM-48. 2) Bulk Et_4NBH_4. 3) Methyl MCM-48 loaded with Et_4NBH_4. 4) Sample 3 after 200°C in vac. 2hrs. 5) Sample 4 after 300°C in vac. 2hrs.

Decomposition of the $[Et_4N]^+$ cation and $[BH_4]^-$ anion are simultaneous as seen with adsorbeded Me_4NBH_4. Notably, the surface Si-H bonds formed during the decomposition of Me_4NBH_4 do not form during the decomposition of adsorbed Et_4NBH_4 despite the significant release of hydrogen. This would imply that the surface Si-H bond formed during the decomposition of the Me_4NBH_4 is a part of the decomposition mechanism, rather than a subsequent formation from the produced gaseous hydrogen.

ACKNOWLEDGMENTS

This work has benefited from the use of FDS at the Lujan Center at Los Alamos Neutron Science Center, funded by DOE Office of Basic Energy Sciences. Los Alamos National Laboratory is operated by Los Alamos National Security LLC under DOE Contract DE-AC52-06NA25396.

REFERENCES

1. Bogdanovic, B.; Felderhoff, M.; Streukens, G. *J. Serb. Chem. Soc.* 72, 183-196, 2009.
2. Zuttel, A.; Wenger, P.; Rentsch, S.; Sudan, P.; Mauron P.; Emmenegger, C. *J. of Power Sources* **118**, 1-7, 2003.
3. DOE, BES *avail. at http://www.sc.doe.gov/bes/hydrogen.pdf* **2004**.
4. Ozolins, V.; Majzoub, E. H.; Wolverton, C. *J. Am. Chem. Soc.* 2009, **131**, 230.
5. Siegel, D.; Wolverton, C.; Ozolins, V. *Phys. Rev. B.* **76**, 134102, 2007.
6. Zaluska, A.; Zaluski, L.; Strom-Olsen, J. O. *J. Alloys Compounds,* **298**, 125, 2000.
7. Fang, Z.; Wang P.; Rufford, T.; Kang, X.; Lu, G.; Cheng, H. *Acta Mat.* **56**, 6257, 2008.
8. Romero, A.; Alba, M.; Zhou, W.; Klinowski, J.; *J. Phys. Chem. B* **101**, 5294, 1997.
9. Mitzutani, G.; Ushioda, S. *J. Chem. Phys.* **91**(1), 598, 1999.
10. Eckert, J.; Sewell, T.; Kress, J.; Kober, E.; Wang, L.; Olah, G.; *J. Phys. Chem. A* **108**, 11369, 2004.

Mater. Res. Soc. Symp. Proc. Vol. 1262 © 2010 Materials Research Society   1262-W03-04

NaBX$_4$-MgX$_2$ Composites (X= D, H) Investigated by *In Situ* Neutron Diffraction

D. Pottmaier[1], S. Garroni[2], M. Brunelli[3], G. B. M. Vaughan[4], A. Castellero[1], E. Menéndez[2], M.D. Baró[2], and M. Baricco[1]

[1]Dipartimento di Chimica IFM - NIS - Università di Torino - Turin, Italy;
[2] Departament de Física - Universitat Autònoma de Barcelona - Barcelona, Spain;
[3] D20 - Institut Laue-Langevin - Grenoble, France;
[4] ID11 - European Synchrotron Radiation Facility - Grenoble, France.

ABSTRACT

Light element complex hydrides (e.g. NaBH$_4$) together with metal hydrides (e.g. MgH$_2$) are considered two primary classes of solid state hydrogen storage materials. In spite of drawbacks such as unfavorable thermodynamics and poor kinetics, enhancements may occur in reactive hydride composites by nanostructuring of reactant phases and formation of more stable product phases (e.g. MgB$_2$) which lower overall reaction enthalpy and allow reversibility. One potential system is based on mixing NaBH$_4$ and MgH$_2$ and subsequent ball milling, which in a 2:1 molar ratio can store considerable amounts of hydrogen by weight (up to 7.8 wt%). A study of the 2NaBX$_4$ + MgX$_2$ → MgB$_2$ + 2NaX + 4X$_2$ (X=D,H) reaction is assessed by means of in-situ neutron diffraction with different combinations of hydrogen and deuterium in the X position. The desorption is established to begin at temperatures as low as 250 °C, starting with decomposition of nanostructured MgX$_2$ due to joint effects of nanostructured MgX$_2$ and its reducing effect at NaBX$_4$. Analyses of background profile, due to the high incoherent neutron scattering of hydrogen, as a function of temperature demonstrate direct correlation of H/D desorption reactions with relative phases amount.

INTRODUCTION

Hydrogen gas has the highest heating value among all chemical fuels, but it has also low energy content per unit volume [1]. Therefore, finding an alternative way to store hydrogen is of critical importance for the implementation of an efficient hydrogen energy system. Hydrogen can be stored in the solid state in two modes: in ad/physorption - hydrogen remains in its molecular form interacting with compounds (e.g. Carbon polymorphs) by weak bonds; whereas in ab/chemisorption - hydrogen in its atomic form reacts with other elements (e.g. Alkali metals) by forming primary bonds. With its high hydrogen storage capacity of 10.8 wt%, NaBH$_4$ is an excellent candidate for solid hydrogen storage applications, but its high thermal stability and lack of reversibility hinder its use. Enhancements may occur in reactive hydride composites (RHC), a novel approach consisting on mixing a complex hydride (e.g. NaBH$_4$) with a single hydride (e.g. MgH$_2$) which lowers overall reaction enthalpy and allows reversibility of this hydrogen storage composite system [2]. Significantly, improved ab- and desorption kinetics are achieved through RHC systems compared to single complex hydrides, while maintaining high hydrogen storage capacities. For NaBH$_4$-MgH$_2$ system, possible decomposition reactions are the following:

$$2NaBH_4 + MgH_2 \leftarrow \rightarrow 2NaH + MgB_2 + 4H_2 \text{ [7.8 wt%]} \quad (1)$$
$$2NaBH_4 + MgH_2 \leftarrow \rightarrow 2Na + MgB_2 + 5H_2 \text{ [9.8 wt%]} \quad (2)$$

Although RHCs thermodynamic properties have been explained by MgB_2 formation [3, 4], no full understanding of the structural and microstructural developments during desorption and absorption reactions has not been reached yet. *Ex situ* X-ray diffraction (XRD) of the hydrogenated RHC system shows two unidentified diffraction peaks [4]. In addition, the existence of intermediate compounds during desorption of metal borohydrides has been suggested [5]. For $NaBH_4$, a hydrogen release has been reported at low temperatures, well below the melting point of the pure compound [6]. *Ex situ* XRD of the RHC system after heat treatment at different temperatures has already been performed in the laboratory [7], and a preliminary *in situ* synchrotron diffraction study [8] has shown that the reaction between $NaBH_4$ and MgH_2 does indeed lower the dehydrogenation temperature. Even though these experiments imply a possible role of an intermediate phase, its origin and its role on the reaction mechanism could not be elucidated from these data. In particular for this RHC system, neither the possible precipitation of metallic sodium during hydrogen desorption, nor details of the hydrogen atoms positions can be assessed from these data.

The nature of neutrons interaction with matter makes them an ideal probe source for structural studies. Namely, since neutrons are scattered by nuclei instead of electron shells, as occurs with photons (X-rays), neutrons have high sensitivity to many elements with low atomic number. Furthermore, neutrons can penetrate deeply into certain materials (e.g. stainless steel) becoming especially important for experiments carried out under controlled atmosphere, such as hydrogen sorption reactions. In this work *in situ* neutron diffraction (ND) studies were performed in the $NaBX_4$-MgX_2 (X=D, H) system in order to obtain additional data and better elucidate the reaction mechanisms of this RHC. Particularly, this study provides a description of structural rearrangements due to hydrogen/deuterium release during annealing. It is also discussed which final phases among the expected (NaX+ MgB_2/ Na + MgB_2) are the most favorable.

EXPERIMENT

Composite $NaBX_4$-MgX_2 samples exchanging X position for H or D, with molar ratio 2:1 were prepared from single compound powders by reactive ball milling (BM), i.e. a D-H sample ($2NaBD_4$+MgH_2) and an H-D sample ($2NaBH_4$+MgD_2). RHC samples were ball milled in a Spex Mix/mill 8000 device for 10 hours at 870 rpm. The powders were sealed in a stainless steel vial under Argon atmosphere. The milling process was performed with 2 balls of 12 mm diameter (7 g) each one and 3 g of sample powder per batch. $NaBH_4$, MgH_2, and $NaBD_4$ were purchased from Sigma Aldrich, while MgD_2 was prepared at our laboratory using a Sieverts apparatus (see appendix). Thermal Programmed Desorption (TPD) was performed in continuous mode up to 600°C at a heating rate of 5°C/min with a helium flow of 50 mL/min. MgH2 was used as calibration standard for quantification of hydrogen release, assuming single reaction MgH2 Mg+H2 (7.66 wt%). Inside an argon glove-box samples were loaded into stainless steel (ss) tubes and analyzed using a home made ss apparatus. The default vanadium can inside quartz tube sample holder is unsuitable for this kind of analyses, because of its interaction with hydrogen. ND investigation was performed in the D20 high intensity two-axis diffractometer at the Institut Laue-Langevin (ILL), in Grenoble (F). It was used the high resolution ($\Delta d/d \cdot 10^{-3}$= 3) configuration with Ge (117) monochromator, beam dimension of 50 mm height and 30 mm width, neutron wavelength of 1.3 Å and a flux on the sample of 10^7 s^{-1} cm^{-2}. ND patterns were collected in 1 minute interval and converted using the LAMP (Large Array Manipulation) program. Assessment of structural information contained in the powder ND patterns was calculated using MAUD (Material Analysis Using Diffraction), a general diffraction/reflectivity analysis program mainly based on the Rietveld method and oriented to materials science studies [9].

RESULTS AND DISCUSSION

Collection of neutron diffraction data was performed in the temperature range from room temperature up to 600 °C at 2°C/min and subsequent ambient cooling under continuous vacuum. *In situ* ND patterns collected at selected temperatures of D-H and H-D samples are shown in figure 1.

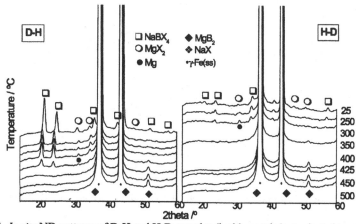

Figure 1. *In situ* ND patterns of D-H and H-D samples (inside ss tube) at selected temperatures.

The ND pattern of both RHC samples (D-H and H-D) show main peaks assigned to the starting phases (NaBX$_4$ and MgX$_2$) and the two major peaks of the Fe-based sample holder (i.e. stainless steel). As expected, whereas the 2 theta positions of the peaks are the same in both hydrogenated and deuteried phases with only relative intensities changed when hydrogen is substituted by deuterium.

ND measurements at room temperature were recorded for approximately 1 hour which results in a total of 50 patterns for each sample. Multiple curve average of these ND patterns at room temperature were used to obtain structural information through Rietveld refinement. Even though the BM process does not induce the formation of new phases (i.e., only presence of the initial NaBX$_4$ and MgX$_2$ phases), it leads to a pronounced nanostructuring effect of MgX$_2$ phase as it can be seen in the broadening of its peaks. RHCs are usually prepared under controlled atmosphere by BM, a mechanical process which has a different influence depending on the nature of the compounds. The effects of BM on single compounds (NaBH$_4$, MgH$_2$) have been identified separately by *ex situ* XRD measurements at room temperature [10, 11]. While NaBH$_4$

shows high structural stability under BM conditions [10], MgH_2 is crushed into nanostructured particles with a bimodal size distribution [11].

In order to estimate possible H/D substitution during BM observed by Infrared spectroscopy [12], the experimental ND pattern of the as-milled D-H sample is compared with calculated patterns according to different structural models. These results, experimental data and calculated patterns, are shown in figure 2 together with the residuals. A possible exchange between BH_4 and BD_4 units in $NaBX_4$ has been estimated by refining of ND pattern with suitable amounts of both $NaBH_4$ and $NaBD_4$ phases. Moreover, possible exchanges between H and D atoms in the tetrahedron have been estimated refining the H and D sites occupancy in the structures. The fittings show a limited H/D exchange during BM, as evidenced by the small variation in the phases amount and in the sites occupancy reported in the table of figure 2.

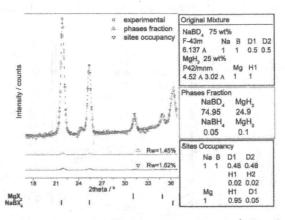

Figure 2. Rietveld refinement and structural details for the as-prepared D-H sample.

The first change in the ND patterns occurs in the heating ramp up to 250 °C, when the MgX_2 phase vanishes and the pure Mg phase start to form for both D-H and H-D samples. As the temperature is increased, main peaks of the $NaBX_4$ and Mg phases gradually fall into the background up to 400 °C when the MgB_2 phase starts to rise from the background. It is believed that MgB_2 may be formed and it is already present at lower temperatures, but it was either too dispersed and/or nanocrystalline to give identifiable ND peaks. Furthermore, the absence of significant peaks of any Na-based phase in the ND diffractogram suggests a desorption reaction resulting in single Na phase instead of a NaX (NaH, NaD) phase, justified by the fact that pure Na is in its liquid state in this temperature range.

ND intensities are attenuated due to incoherent scattering of hydrogen in the structure, hydrogen having the largest incoherent scattering interaction among all the elements. For this reason, ND patterns have different background intensities depending on the hydrogen content of the starting sample and as a function of temperature. Therefore, analysis of the background profile might be a reference to the presence of hydrogenated/deuterated phases along the desorption reaction. Values of background intensity for different samples as a function of temperature are reported in figure 3. The ND background is considered to be the intensity between peaks belonging to crystalline phases. In this work, this was taken at the 18.5° 2 theta position. Obtaining a semi-quantitative measure of hydrogen content based on background intensity is based on some assumptions. These consider the uniformity of sample amount and compacting factor, and most importantly in the uniformity on neutrons diffracting power through the sample.

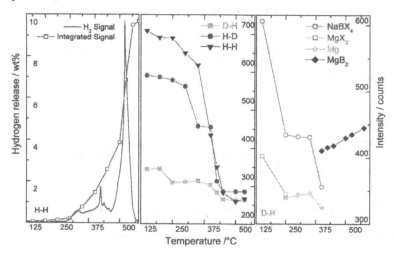

Figure 3. Hydrogen release profile (left), background (center), and peaks (right) intensity as a function of temperature.

In order to confirm the total hydrogen release from the RHC system during heat treatment, TPD results are shown in figure 3 (left panel). During annealing at 5 °C/min, H-H sample shows a complete desorption of about 10 wt% up to 550 °C, in agreement with desorption reaction (2).

As can be observed in figure 3 (center panel), background intensities of D-H sample show a pronounced decrease at 250 °C which corresponds to the decomposition of MgH_2. Conversely, background intensities of H-D sample evidence no significant change with MgD_2 decomposition. Moreover, the observed decrease in the background intensity with temperature proves the main decomposition of the $NaBH_4$ phase during heating at temperatures up to 400 °C. This result also confirms the formation of a nanostructured and dispersed MgB_2 phase. However, complete decomposition of $NaBX_4$ and formation of MgB_2 is reached only at 450 °C as from this point background intensities are constant.

The obtained results are confirmed by the analysis of the phase evolution during annealing. The peak intensities of the various phases observed as a function of temperature in the D-H composite are reported in figure 3 (right panel) as a function of temperature. The decrease of the intensity of the $NaBX_4$ and MgX_2 phases is evident at low temperatures. In the temperature range between 250 °C and 350 °C, Mg is formed as a decomposition product of MgX_2. At higher temperatures, the reaction between $NaBX_4$ and Mg leads to the formation of MgB_2.

CONCLUSIONS

ND patterns and in-situ studies of $NaBX_4$-MgX_2 composites were shown in this work. These have mainly confirmed that desorption reaction in vacuum of this RHC follow a different trend to other MgH_2 composite systems. Thus, it is most likely to start already at 250 °C with MgH_2 decomposition to single Mg phase. The pure Mg destabilizes the complex hydride $NaBH_4$ leaving pure Na liquid phase and a dispersed-nanocrystalline MgB_2 phase. From 400°C, the presence of MgB_2 phase is evident on ND pattern. Further studies are under development to identify mechanisms and phases induced by a hydrogen pressure in determining reaction pathway for the $NaBH_4$-MgH_2 system; with the possibility of passing through intermediate phases (e.g. $NaMgH_3$).

APPENDIX

MgD_2 compound was synthesized through first decomposition of MgH_2 using a heating rate of 2 °C/min up to 425 °C under 1 atm in 3 hours, and subsequent deuteration of Mg (14.2 wt%) during 24 hours at 350 °C under 50 atm.

ACKNOWLEDGMENTS

A special thanks to Alain Daramsy (ILL). This study was financially supported by the European Commission under FP6 project COSY (MRTN-CT-2006-035366). E. M. thanks the Fund for Scientific Research – Flanders (FWO) for financial support.

REFERENCES

1. A. Zuettel, A. Borgschulte, L. Schlapbach, *Hydrogen as Future Energy Carrier*, (Wiley-VCH, Weinheim, 2008) pp. 91.

2. G. Barkhordarian, T. Klassen, R. Bormann, International Patent No. WO2006/063627.

3. J.J. Vajo, F. Mertens, C. C. Ahn, R. C. Bowman, Jr., B. Fultz, J. Phys. Chem. B 108, 13977-13983 (2004).

4. G. Barkhordarian, T. Klassen, M. Dornheim, R. Bormann, J. Alloys Compd. 440, L18-L21 (2007).

5. S.-J. Hwang, R.C. Bowman, Jr., J.W. Reiter, J. Rijssenbeek, G.L. Soloveichik, J.-C. Zhao, H. Kabbour, C.C. Ahn, J. Phys. Chem. C 112, 3164-3169 (2008).

6. J. Urgnani, F.J. Torres, M. Palumbo, M. Baricco, Int. J. Hydrogen Energy 33, 3111-15 (2008).

7. D. Pottmaier, E. Groppo, S. Bordiga, G. Spoto, M. Baricco, *Dehydrogenation pathway of 2NaBH_4+MgH_2 composite*, (HYSYDAYS 2009 Proc., Turin, Italy, 2009).

8. S. Garroni, C. Pistidda, M. Brunelli, G.B.M. Vaughan, S. Suriñach, M.D. Baró, Scripta Mater. 60, 1129-1134 (2009).

9. L. Lutterotti, S. Matthies, H.-R. Wenk, A. J. Schulz, J. Richardon, J. of Apply. Phys. 81, 594-600 (1997). *MAUD is available at http://www.ing.unitn.it/~maud/*

10. R.A. Varin, Ch. Chiu, J. Alloys Compd. 397, 276-281 (2005).

11. J. Huot, G. Liang, S. Boily, A. Van Neste, R. Schulz, J. Alloys Compd. 293-295, 495-500 (1999).

12. D. Pottmaier, M. Baricco, presented at the 7th COSY Project Meeting, Grenoble, France, 2009 (unpublished).

Mater. Res. Soc. Symp. Proc. Vol. 1262 © 2010 Materials Research Society   1262-W03-05

Neutron Scattering to Characterize Cu/Mg(Li) Destabilized Hydrogen Storage Materials

M.H. Braga[1,2], M. Wolverton[1], A. Llobet[1], and L.L. Daemen[1]

[1] LANSCE-LC, Los Alamos National Laboratory, mail stop: H805, NM, 87545, USA.

[2] CEMUC, Engineering Physics Department, Porto University - FEUP, R. Dr. Roberto Frias, s/n, 4200-465, Porto, Portugal.

ABSTRACT

Cu-Li-Mg-(H,D) was studied as an example of destabilizer of the Ti-(H,D) system. A Cu-Li-Mg alloy was prepared resulting in the formation of a system with 60.5 at% of $CuLi_{0.08}Mg_{1.92}$, 23.9 at% of $CuMg_2$ and 15.6 at% of Cu_2Mg. Titanium was added to a fraction of this mixture so that 68.2 at% (47.3 wt%) of the final mixture was Ti. The mixture was ground and kept at 200 °C/473 K for 7h under H_2 or 9h under D_2 at P = 34 bar. Under those conditions, neutron powder diffraction shows the formation of TiD_2, as well as of the deuteride of $CuLi_{0.08}Mg_{1.92}$. Similarly inelastic neutron scattering shows that at 10 K TiH_2 is present in the sample, together with the hydride of $CuLi_{0.08}Mg_{1.92}$. Interestingly, at 10 K TiH_2 is very clearly detected and at 300 K TiH_2 is still clearly present as indicated by the neutron vibrational spectrum, but $CuLi_{0.08}Mg_{1.92}$-H is not detected anymore. These results indicate that $Ti(H,D)_2$ is possibly formed by diffusion of hydrogen from the Cu-Li-Mg-(H,D) alloys. This is an intriguing result since TiH_2 is normally synthesized from the metal at T > 400°C/673 K (and most commonly at T ~ 700 °C/973 K). In the presence of $CuLi_{0.08}Mg_{1.92}$, TiH_2 forms at a temperature that is 300 – 400 K lower than that needed to synthesize it just from the elements.

INTRODUCTION

The pioneering work of Reilly and Wiswall [1] on hydrogen storage in $CuMg_2$ provides the first clear example of destabilization. $CuMg_2$ was reversibly hydrogenated to $3/2MgH_2 + 1/2Cu_2Mg$ with an equilibrium pressure of 1 bar at 240 °C/513 K. This temperature is ~ 40 K lower than T(1 bar) for pure MgH_2. However, since $CuMg_2$ does not form a hydride, this work was set aside until very recently.

The current search for an on-board hydrogen storage material has led to a point where only a system of 4 elements will cover all specifications and this with difficulty. A destabilization strategy then becomes a viable, attractive path forward for further progress.

$CuMg_2$ has an orthorhombic crystal structure (Fddd). However $CuLi_xMg_{2-x}$ (x = 0.08) has a hexagonal crystal structure (P6$_2$22), just like $NiMg_2$ - a compound known for its hydrogen storage properties. $NiMg_2$ absorbs up to 3.6 wt% of hydrogen, at 1 bar and 282 °C/555 K. In spite of the fact that the percentage of H_2 absorbed by $NiMg_2$ is enough to propitiate practical applications, the temperature at which the alloy desorbs hydrogen is much too high for current applications. Still, the alloy can be found in practical applications when added to other elements/alloys. A comparison between the phase diagrams of the systems Cu-Mg and Ni-Mg shows that these binary systems form compounds with similar stoichiometry. $NiMg_2$ is formed by peritectic reaction of the elements at 759 °C/1032 K and $CuMg_2$ at 568 °C/841 K by congruent melting. The presence of Li lowers even further the melting point of $CuMg_2$. Since the energy of formation of the hydride is related to that of the primary alloy, it was hypothesized that $CuLi_xMg_{2-x}$ might also be a hydrogen storage material similar to $NiMg_2$. Presumably, its advantage would be that it would release hydrogen at a lower temperature (possibly close to room temperature). Preliminary studies at the Manuel Lujan, Jr. Neutron Scattering Center showed that $CuLi_xMg_{2-x}$ absorbs 5.3 wt% H at an equilibrium pressure of approx. 27 bar at 200 °C/473 K. DSC/TG experiments show that a considerable amount of hydrogen can be released at T < 100 °C/373 K. If these results are confirmed, this will mean that, not only $CuLi_xMg_{2-x}$ absorbs a considerable amount of hydrogen, but also will probably release it at a temperature in the range of 50 °C /323 K to 200 °C/473 K, where applications are easier to develop. Hence it should be possible to use this alloy with fuel cells or in batteries. It was also observed that a sample containing $CuMg_2$ could release hydrogen at 180 °C/453 K $\leq$ T $\leq$ 210 °C/483 K, probably meaning that the presence of $CuLi_xMg_{2-x}$ will make MgH_2 releasing hydrogen at an even lower temperature. In this work we have characterized Cu-Li-Mg+Ti+(H,D) hydrogen storage systems and its thermodynamic properties by means of neutron scattering and other complementary techniques.

EXPERIMENT

An alloy of the system Cu-Li-Mg was prepared by melting the elements together: Cu (electrolytic, 99.99% purity, 325 mesh), Mg (99.8% purity, 200 mesh, Alfa Aesar), and small (less than 3mm wide) pieces of Li (99% purity, Alfa Aesar), at 850 °C/1123 K for 1h using a stirring device. The alloy was quenched into liquid nitrogen. The resulting samples had 60.5 at% of $CuLi_{0.08}Mg_{1.92}$, 23.9 at% of $CuMg_2$ and 15.6 at% of Cu_2Mg. The samples were first characterized by means of XRD using a Rigaku Ultima III powder diffractometer, and their composition was roughly determined by means of the Match software, [2] which uses the "Reference Intensity Ratio method" RiR - method) [3] to obtain phase fractions. Patterns were collected with $CuK\alpha$ typically from $2\theta = 15$ to $70°$ with steps of $0.02°$ and a counting time of 10 s per bin.

This brittle material was mixed with Ti (99.9% purity, 325 mesh, Alfa Aesar) so that 68.2 at%/47.3 wt% of the final mixture was Ti. The mixture was ball-milled for 3 h in a dry box under a He protective atmosphere. Part of the resulting mixture was then divided into two portions and each of them was sealed inside a stainless steel crucible and kept at 200 °C for 7h under H_2 or 9h under D_2 both at P = 34 bar. These samples were then cooled to 5 K (HIPD, neutron powder diffraction) and 10 K (FDS, neutron vibrational spectroscopy) over a period of 2 to 3 hours. Differential Scanning Calorimetry (DSC) / Thermo Gravimetric (TG) measurements were performed under Ar with a Netzsch instrument (STA 449C) from room temperature to 450 °C/723 K, using alumina pans and lids and different heating rates (5 and 10 °C/min.).

Time-of-flight (TOF) neutron diffraction data were collected on the neutron powder diffractometer (HIPD) at the Manuel Lujan, Jr. Neutron Scattering Center at Los Alamos National Laboratory. We have measured a sample containing Cu-Li-Mg-Ti-D at 5 K, 60 K, 100 K, 200 K and 300 K for 8 h / each. The structural data were refined using GSAS [4]. Since there were up to five phases to refine, and since all the phases have well defined structural parameters at all temperatures [5], except the deuteride of $CuLi_{0.08}Mg_{1.92}$ whose structural parameters and stoichiometric composition are not well established yet, but were, nevertheless, Rietveld refined in a previous work [6] (Fig. 1), we have used the Model-biased method [4] for all phases except TiD$_2$ (tetragonal and cubic) [7,8] for which Rietveld refinement was used in the last step to confirm results.

The Filter Difference Spectrometer (FDS) was used for neutron vibrational spectroscopy. We have analyzed, at 10 K, samples in the systems Cu-Li-Mg, Cu-Li-Mg-H, Cu-Li-Mg-Ti, and Cu-Li-Mg-Ti-H. The latter was also analyzed at 300 K.

DISCUSSION

Results from neutron diffraction indicate that from 5 K to 100 K only CuLi$_{0.08}$Mg$_{1.92}$, CuMg$_2$, Cu$_2$Mg, Ti and the deuteride of CuLi$_{0.08}$Mg$_{1.92}$ that was refined as being monoclinic C 2/m as in previous work shown in Fig. 1 were present. At 200 K the system can be described as having the same phases that at 100 K, except for the fact that instead of Ti, there are new peaks corresponding to TiD$_2$ (tetragonal) (Figs. 2, 3). At 300 K, TiD$_2$ suffers a structural phase transition and it is no longer tetragonal but cubic (Figs. 2, 3) as reported before in the literature [7]. These results support the hypothesis that the Cu-Li-Mg system is the first phase to become deuterised. When the sample containing Cu-Li-Mg-Ti-D was quenched to 5 K, after absorbing D$_2$ at 200 °C/473 K, it could not be noticed the presence of any of Ti's deuteride phases but it was possible to observe the presence of the deuteride phase of CuLi$_x$Mg$_{2-x}$ (x = 0.08). Then, it seems that by diffusion, Ti will also become deuterised above 100 K (possibly near 200 K).

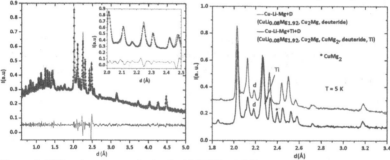

Figure 1. Diffraction pattern obtained with HIPD, at 5 K of a sample containing 75.6 at% of CuLi$_{0.08}$Mg$_{1.92}$, 24.4 at% of Cu$_2$Mg and that was deuterised at 200 °C/473 K. Rietveld refinement shows that the deuteride phase of CuLi$_{0.08}$Mg$_{1.92}$, CuLi$_{0.08}$Mg$_{1.92}$D$_4$, can be monoclinic C 2/m. Nevertheless, these results are still preliminary. On the right, comparison between the diffraction patterns of samples containing Cu-Li-Mg+D (the same pattern as on the left) and Cu-Li-Mg+D+Ti.

From FDS measurements, it is possible to clearly detect the presence of TiH$_2$ (Fig. 3 – right), since 10 K, by the presence of a peak at 1155 cm^{-1} corresponding to vibration of the H in the Ti positions, Ti-H. At this temperature it is still present the hydride phase of CuLi$_{0.08}$Mg$_{1.82}$ (noticed by the presence of bending modes correspondent to the peaks at around 470, 592 and 725 cm^{-1} that can be seen both on the Cu-Li-Mg-H and Cu-Li-Mg-Ti-H (10K) samples), but at

300 K, this phase cannot be seen anymore. Thus, the hypothesis about hydrogen diffusing from the Cu-Li-Mg-H system to Ti can also be supported by these measurements as well.

Analyzing DSC curves, it can be inferred from Fig. 4 that after FDS measurements, there was only 0.3 wt% of mass loss from room temperature to 450 °C/723 K. Mass loss started before 100 °C/373 K, as it happens in the Cu-Li-Mg-H system, and it is probably due to hydrogen desorption from $CuLi_{0.08}Mg_{1.92}$-H. On the other hand, it cannot be seen a new rate of mass loss after approx. 200 °C/ 473 K, as it can be seen in Fig. 5 for the Cu-Mg-Li-H samples. The peak correspondent to 282 °C/555 K it is not observable in the DSC curves of Fig. 4 as well. This is probably due to the fact that instead of $CuLi_{0.08}Mg_{1.92}$-H destabilizing MgH_2 that is formed from $2CuMg_2 + 3H_2 \leftrightarrow 3MgH_2 + Cu_2Mg$, in presence of Ti, it will destabilize or act as a catalyst of Ti.

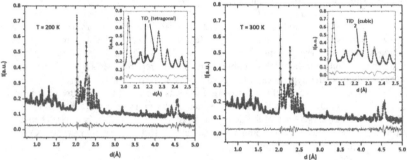

Figure 2. Diffraction patterns obtained with HIPD, at 200 K and 300 K of a sample containing 31.8 % of (60.5 at% of $CuLi_{0.08}Mg_{1.92}$, 23.9 at% of $CuMg_2$ and 15.6 at% of Cu_2Mg) + 68.2 at% of Ti and that was deuterised at 200 °C/473 K. Rietveld refinement shows the tetragonal structure of TiD_2 at 200 K and the cubic at 300 K.

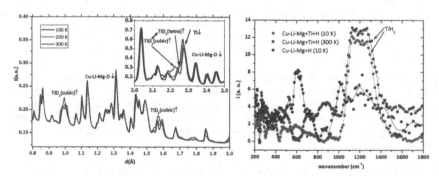

Figure 3. Diffraction patterns and inelastic spectra, obtained at FDS, of a sample containing 31.8 % of (60.5 at% of $CuLi_{0.08}Mg_{1.92}$, 23.9 at% of $CuMg_2$ and 15.6 at% of Cu_2Mg) + 68.2 at% of Ti and that was deuterised (diffraction)/hydrogenised(spectroscopy) at 200 °C/473 K. Results show the presence of TiD_2/TiH_2.

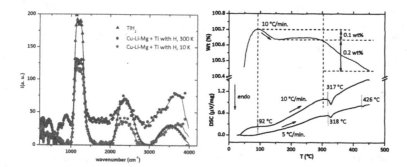

Figure 4. Inelastic spectra of a sample containing 31.8 % of (60.5 at% of $CuLi_{0.08}Mg_{1.92}$, 23.9 at% of $CuMg_2$ and 15.6 at% of Cu_2Mg) + 68.2 at% of Ti and that was hydrogenised at 200 °C/473 K in comparison with the TiH_2 spectrum. Results show the presence of TiD_2/TiH_2. DSC/TG curves of a sample containing 31.8 % of (60.5 at% of $CuLi_{0.08}Mg_{1.92}$, 23.9 at% of $CuMg_2$ and 15.6 at% of Cu_2Mg) + 68.2 at% of Ti and that was hydrogenised at 200 °C/473 K after being measured in FDS.

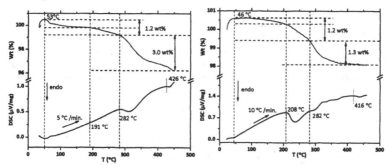

Figure 5. DSC/TG curve of a sample containing 60.5 at% of $CuLi_{0.08}Mg_{1.92}$, 23.9 at% of $CuMg_2$ and 15.6 at% of Cu_2Mg and that was hydrogenised at 200 °C/473 K and of a sample containing 64.3 at% of $CuLi_{0.08}Mg_{1.92}$, 19.6 at% of $CuMg_2$ and 16.1 at% of Cu_2Mg and that was also hydrogenised at 200 °C/473 K.

Hence, 317 °C/590 K will probably be lower limit temperature for TiH_2 to release hydrogen.

Another important support for the previous hypothesis is that neither in the neutron diffraction patterns, nor in FDS we could detect new phases. Further, at 416 °C /689 K-426 °C/699 K, there is a clear endothermic peak corresponding to the melting point of $CuLi_{0.08}Mg_{1.92}$ indicating the previous presence of this phase.

The amount of mass loss is not conclusive about the stoichiometry of the hydrides, since DSC/TG is not the suitable measurement to obtain this information, and further we did not expect that the samples were saturated with hydrogen prior to the experiment. Additionally, samples were not cycled before and the involving atmosphere was not of H_2 as it would be more appropriate. Nonetheless, these experiments give valuable information about approximated temperatures at which the system starts to lose mass at a certain rate.

It is also possible that additionally to the mass loss effect; there will be a slightly increase of mass due to oxidation, although this effect is covered by the first.

CONCLUSIONS

We have added Ti to a system composed by Cu-Li-Mg and the mixture was hydrogenated/deuterated. Neutron scattering, XRD and DSC experiments were performed and it can be concluded that, in the presence of Ti mechanically mixed, $CuLi_{0.08}Mg_{1.92}$ will destabilize or catalyze Ti and $Ti(H,D)_2$ will be synthesized at 200 K. The latter represents a very significant decrease of temperature since $Ti(H,D)_2$ can only be synthesized from Ti at T > 773 K, usually at 1000 K. On the other hand, it is possible that TiH_2 will start releasing hydrogen (at 1 bar) at ~ 590 K which is also a significant decrease from 700 K - the lowest value found in the literature for the latter to occur.

ACKNOWLEDGMENTS

MHB and LLD would like to acknowledge Portuguese Science Foundation, FCT, for the project (PTDC/CTM/099461/2008 and FCOMP-01-0124-FEDER-009369). This work has benefited from the use of HIPD and FDS at the Lujan Center at Los Alamos Neutron Science Center, funded by DOE Office of Basic Energy Sciences. Los Alamos National Laboratory is operated by Los Alamos National Security LLC under DOE Contract DE-AC52-06NA25396.

REFERENCES

1. J.J. Reilly, R.H. Wiswall, Inorg. Chem., 6(12) (1967) 2220-2223.
2. Match, http://www.crystalimpact.com/, 2009.
3. P.M. de Wolff, J.W. Visser, Absolute Intensities. Report 641.109. Technisch Physische Dienst, Delft, Netherlands. Reprinted (1988) Powder Diffract 3:202-204.
4. A.C. Larson, R.B. von Dreele, GSAS Generalized Structure Analysis System, LANSCE, Los Alamos, 2004.
5. M.H. Braga, J.J.A. Ferreira, J. Siewenie, T. Proffen, S.C. Vogel, L.L. Daemen, J. of Sol. Stat. Chem., 183(1) (2010) 10-19.
6. M.H. Braga, M. Wolverton, M. Hartl, H. Xu, Y. Zhao, L.L. Daemen, RGSAM – Rio Grande Symposium on Advanced Materials, Albuquerque, NM, USA, October 5th, 2009, P14, p. 27.
7. H.L.jr. Yakel, Acta Crystall. 11 (1958) 46-51.
8. P.E. Irving, C.A. Beevers, Metall. Trans. 2 (1971) 613-615.

Mater. Res. Soc. Symp. Proc. Vol. 1262 © 2010 Materials Research Society   1262-W03-09

Ex-situ and in-situ neutron radiography investigations of the hydrogen uptake of nuclear fuel cladding materials during steam oxidation at 1000°C and above

Mirco Grosse [1], Marius van den Berg [1,2], Eberhard Lehmann [3], and Burkhard Schillinger [4]
[1] Karlsruhe Institute of Technology, Germany; Institute for Materials Research
[2] Fontys University Eindhoven, Technische Natuurkunde, Netherlands
[3] Paul Scherrer Institute Villigen, Switzerland, Department of Spallation Neutron Source
[4] Technical University Munich, ZWE FRM-II, Germany

ABSTRACT

Neutron radiography is a powerful tool for the investigation of the hydrogen uptake of zirconium alloys. It is fast, fully quantitative, non-destructive and provides a spatial resolution of 30 µm. The non-destructive character of neutron radiography provides the possibility of in-situ investigations. The paper describes the calibration of the method and delivers results of ex-situ measurements of the hydrogen concentration distribution after steam oxidation, as well as in-situ experiments of hydrogen diffusion in β-Zr and in-situ investigations of the hydrogen uptake during steam oxidation.

INTRODUCTION

The most important accident management measure to terminate a severe accident transient in a Light Water Reactor (LWR) is the injection of water to cool down the uncovered degraded core. The combination of hot fuel rods and steam results in a strong exothermic oxidation reaction of the zirconium alloys used as cladding material, followed by a sharp increase in temperature, hydrogen production and fission product release. Free protons are produced in the steam oxidation reaction with oxygen vacancies $V_O^{(2+)}$.

$$2\,H_2O + Zr + 2V_O^{(2+)} + 4e^- \rightarrow ZrO_2 + 4p^+ + 4e^- \qquad (1)$$

They can recombine to H_2 gas and are released. Otherwise they can diffuse through the growing oxide layer and be absorbed by the β-Zr phase.

$$4p^+ + 4e^- \rightarrow 2 \cdot x\,H_2 \uparrow + 4(1-x)\,H_{\text{ absorbed}} \qquad (2)$$

Whereas the released hydrogen results in the risk of a hydrogen detonation in the reactor environment, the hydrogen absorption shifts the time scale of the hydrogen release and reduces the toughness of the material.

In order to study the hydrogen uptake of cladding materials, usually hot extraction is applied. This method is destructive, time consuming und does not provide any information about local hydrogen distributions. Systematic errors can occur by hydrogen adsorption in the furnace and at the pipe walls and by reaction of the zirconium oxide layer with the sample holder. Neutron radiography provides a fast, fully quantitative and non-destructive method for the determination of hydrogen in zirconium alloys with a spatial resolution up to 30 µm. The non-destructive character of neutron radiography provides the possibility of in-situ experiments.

In this paper the basics of the quantitative analysis of neutron radiographs are given. The calibration of the method will be described and results of ex-situ investigations of specimens withdrawn from The QUENCH large scale severe accident simulation experiments and of in-situ investigations of the hydrogen diffusion and uptake during steam oxidation will be presented.

QUANTITATIVE ANALYSIS OF NEUTRON RADIOGRAPHS

Theoretical basis

In the neutron radiography experiment a parallel neutron beam passes the specimen and is detected by means of a two-dimensional neutron detector, e.g. a CCD camera based system. Different attenuations of the neutron beam of different sample points results in intensity contrasts. The transmission of the specimen $T(x,y)$ detected at a certain point of the radiograph is given by the ratio of the intensity behind and in front of it, $I(x,y)$ and $I_0(x,y)$, respectively, corrected by the background intensity $I_B(x,y)$.

$$T(x, y) = \frac{I(x, y) - I_B(x, y)}{I_0(x, y) - I_B(x, y)} = \exp\left(-\Sigma_{total}(x, y)\, s(x, y)\right) \qquad (3)$$

It is connected exponentially with the total macroscopic neutron cross section of the specimen $\Sigma_{total}(x,y)$ and with the path length of the neutron through the specimen $s(x,y)$. This relation is valid in first order until multiple scattering effects in the assembly can be ignored. The macroscopic total cross section depends on the microscopic cross section σ of the isotopes included and their number density N in the specimen.

$$\Sigma_{total} = \sum_i N_i \sigma_i = \underbrace{N_{Zr}\sigma_{Zr} + N_{Sn}\sigma_{Sn} + N_{Nb}\sigma_{Nb} + \ldots}_{\Sigma_{total}\ (as\ received)} + N_H\sigma_H + N_O\sigma_O \qquad (4)$$

The total microscopic cross sections of the isotopes depend on their type, on their bonding state in the lattice and in some cases on the temperature. In the case of steam oxidation of zirconium alloys the total macroscopic neutron cross section is the sum of the alloy in the received state, the contribution of absorbed hydrogen and of the oxygen in the oxide scale and absorbed in the α-Zr phase. From equation (4) follows that the total macroscopic of the specimen depends linearly on the H/Zr atomic ratio. Also the effect of oxygen has to be corrected.

Radiography experiments

The ex-situ and most of the in-situ neutron radiography experiments were performed at the ICON facility at the Swiss Neutron Source SINQ at PSI Villigen (Switzerland). Some of the in-situ steam oxidation experiments were performed at the ANTARES facility at the FRM-2 reactor (TU Munich, Germany). For the in-situ experiments a high temperature vertical reaction tube furnace (INRRO) was constructed. This furnace has two beam windows which are transparent for neutrons and located at opposite sites of the reaction tube. It allows annealing in inert, reducing and oxidizing atmospheres including steam at temperatures up to 1700 K. The gas composition can be controlled by Bronkhorst mass and volume flow controllers evaporators and mixers. Because the furnace is water cooled outside the sample area and field of view during the measurements it can also be applied close to temperature sensitive optical components like lenses or mirrors.

In-situ neutron radiography investigations were performed during three beam times. In 11/2008 the device was commissioned at the ICON facility. Illumination times of 118 s per frame were applied for these experiments. In 12/2009 calibration measurements were done with illumination times of 19 s per frame. Because of the high intensity of the neutron source FRM-2, the illumination time per frame during the steam oxidation experiments could be reduced to 7.8 s in 01/2010. A frame sequence of 0.1 s^{-1} was been realised.

Calibration of the correlation of total neutron cross section and H/Zr atomic ratio

In order to calibrate the dependence of the total neutron cross section at room temperature samples were loaded with hydrogen by annealing at different temperatures in Ar/H_2 flowing atmosphere with various hydrogen partial pressures p_{H2}. The correlation between p_{H2}, temperature T and concentration $c_H(metal)$ of the absorbed hydrogen in the metal is given by Sieverts´ law:

$$c_H(metal) = \exp\left(\frac{\Delta S}{R} - \frac{\Delta H}{RT}\right)\sqrt{p_{H_2}} \qquad (5)$$

ΔS and ΔH are the free entropy and enthalpy, respectively, R the universal gas constant $(8.314\,J\,mol^{-1}K^{-1})$. The masses of absorbed hydrogen were determined by balancing the specimens before and after the annealing. The total microscopic cross section of hydrogen is about one order of magnitude higher than the cross section of zirconium. Figure 1 shows the correlation between the H/Zr atomic ratio and the total macroscopic neutron cross section of the calibration specimens for different experimental setups of the ICON facility at SINQ. According to equation (4), a linear dependence was found. A similar calibration was performed for the effect of oxygen by means of specimens annealed in argon/oxygen atmosphere.

The calibration for the in-situ experiments is much more difficult. Hydrogen is at temperatures of 1100 K and above very volatile. From a pre-loaded specimen hydrogen would be released immediately into an inert gas atmosphere. Also the calibration obtained at room temperature where the hydrogen is precipitated as hydrides can not be used for high temperatures. At temperatures of about 800 K and above the β-Zr phase can solve up to 40 at.% hydrogen. Due to the parameters of the Sieverts´ law are well determined [1] a equilibrium between hydrogen partial pressure in the gas atmosphere of the furnace and H concentration in the zirconium alloy can be used for the calibration measurements. The specimens were annealed in an Ar/H_2 atmosphere. The hydrogen concentration was controlled by the gas flow controller. In order to correct the effect of the furnace and the sample holder in the furnace the neutron radiographs were referenced with the frame of the sample at the beginning of the test at test temperature but before starting the hydrogen flow. The resulting image shows only the hydrogen in the sample. From the p_{H2} the concentrations in the metal was calculated using equation (5). Figure 2 shows the dependence of the total macroscopic hydrogen cross section of the hydrogen solute in the β-Zr phase for the temperatures investigated, measured at the ICON facility. A linear dependence of Σ on the H/Zr atomic ratio was found. No temperature dependence of Σ is visible. However, the total neutron cross section of hydrogen solute in the β-Zr is about two times higher than Σ of hydrogen precipitated at room temperature. The deviations in Σ results from different neutron spectra (with and without Be filter and by different handling of the background intensity (background correction was not possible for the image plate).

Ex-situ investigations of the hydrogen uptake during steam oxidation

In ex-situ investigations samples from separate effect tests and from the QUENCH large scale tests were investigated at the ICON facility. Figure 3 gives an example for the results of the investigation of specimens after isothermal steam oxidation at 1373 K. Different behaviour between the two Zr-Sn alloys (Zry-4, Duplex) and the Zr-Nb alloy E110 was found at this temperature. A detailed description of the experiments and the results is given in [2].

The axial distributions of the hydrogen in specimens prepared from the QUENCH-15 large scale bundle test are given in figure 4. Also here strong differences between the materials

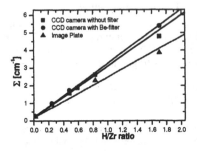
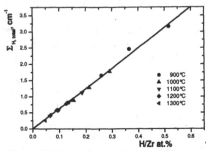

Figure 1 Dependence of the total macroscopic neutron cross section Σ on the H/Zr atomic ratio at room temperature

Figure 2 Dependence of the total macroscopic neutron cross section on the concentration of hydrogen solute in β-Zr at various temperatures

were found. Due to special morphologies of the oxide layer formed during the test, the E110 absorbs about one order of magnitude more hydrogen than Zry-4 or the Westinghouse material ZIRLO. Details of the tests QUENCH-12, -13 and -14 and about the hydrogen distribution determined in specimens withdrawn during the tests or prepared after the tests are given in [3-5].

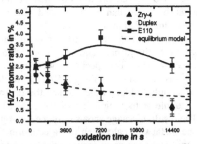
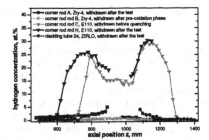

Figure 3 Atomic ratio between hydrogen and zirconium after steam oxidation at 1373 K

Figure 4 Axial hydrogen distribution in specimens prepared from the QUENCH-15 large scale severe accident simulation tests

In-situ investigations of the hydrogen diffusion in β-Zr

In order to study the hydrogen diffusion in β-Zr cylindrical samples with a diameter of 12 mm and a length of 20 mm were prepared from Zry-4. Layers suppressing the hydrogenation were produced on the surface of the specimens by a chemical way. At one end of the specimens the layer was removed by metallographic grinding. Through this basic face of the cylinders the hydrogen can be absorbed. By means of neutron radiography, a sequence of images for the different steps of the diffusion process was taken with a frame frequency of 0.05 and 0.1 s⁻¹, respectively. Figure 5 gives a sequence of radiographs taken with a time step of 600 s during the experiment at 117 K at the ICON facility. The darker the image point, the higher is the Number of protons in the neutron path. The prolongation of the hydrogen diffusion front is well visible. The quantitative analysis of the data is still in progress. The results will be published elsewhere.

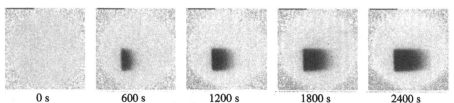

| 0 s | 600 s | 1200 s | 1800 s | 2400 s |

Figure 5 Frame sequence of the hydrogen diffusion into Zry-4 at 1173 K measured at ICON

In-situ investigations of the steam oxidation behaviour of common used cladding materials

The INRRO furnace was also applied to study the process during steam oxidation of Zr alloys in-situ. Hydrogen uptake is important for the mechanical properties of the nuclear fuel cladding tubes. The actual discussion about it shows that the phenomenon is not yet satisfying understood. Ex-situ investigations can not completely answer the open questions like the time dependence of H absorption and release. In-situ neutron radiography investigations provide new possibilities to study this process. Figure 6 gives a frame sequence of the beginning of the steam oxidation for the AREVA Duplex material at 1573 K taken at the ANTARES facility. At t = 0 s the steam injection into the INRRO furnace starts. After 40 s the maximum of c_H (*metal*) in the sample was reached. After this time the hydrogen content in the sample is changed slowly.

Figure 7 gives the time dependence of Σ measured during steam oxidation of Zry-4 at 1273 K (black curve). At this temperature the breakaway effect occurs. Cracks are formed due to a tetragonal to monoclinic phase transition in the oxide layer. During the first 120 s a strong increase in Σ was found. After this rapid increase, Σ changes only slightly. At about 1800 s the breakaway effect starts with the consequence of an additional increase of the neutron cross section. Maximal values of Σ are reached between 2.5 and 3 h. Later it decreases. Additionally to the measured cross section, the estimated contribution of oxygen is given (blue curve). It is calculated from the known oxidation behaviour and from the calibration of the oxygen contribution to Σ. At the end of the experiment parts of the oxide layer spalls. It can not be calculated certainly. Therefore the values of the oxygen effect at times beyond 18000 s are uncertainly. The difference between measured curve and curve describing the effect of oxygen gives the total macroscopic cross section of hydrogen (magenta curve). From this curve the hydrogen concentration was estimated. The first maximum of c_H (*metal*) (~ 250 wppm) is reached during the first frame (illumination time 118 s). A slightly decrease of the hydrogen content follows. When the breakaway effect starts, the hydrogen concentration increases strongly to about 7 to 8 times of the concentration found for compact oxide layers. A detailed discussion of the results obtained by the in-situ neutron radiography experiments will be given elsewhere.

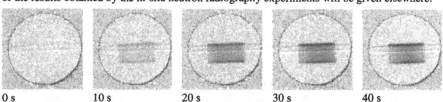

| 0 s | 10 s | 20 s | 30 s | 40 s |

Figure 6 Frame sequence of the beginning of the steam oxidation of the AREVA DUPLEX material at 1573 K

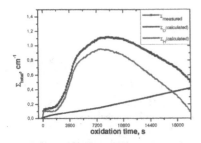

Figure 7 Total macroscopic neutron cross section measured at ICON and estimated for oxygen and hydrogen

CONCLUSIONS

Neutron radiography is a powerful tool to determine quantitatively hydrogen in Zr alloys with high spatial and time resolution. The non-destructive character of the neutron radiography provides the possibility of in-situ investigations of the hydrogen uptake during steam oxidation and the diffusion of the hydrogen. The results of these new experimental possibilities give completely new information about the time dependence of hydrogen uptake and release.

The correlation between hydrogen concentration and total macroscopic neutron cross section were determined for room temperature as well as for temperatures between 1173 and 1573 K. The theoretically known linear dependency of Σ on the atomic ratio between hydrogen and zirconium was experimentally confirmed in our study..

The hydrogen concentration reaches its maximal values after less than one minute. Later, hydrogen is slightly released by the sample. If the breakaway effect occurs, the hydrogen concentration increases too much higher values.

ACKNOWLEDGEMENT

The authors thank all who take part in the preparation of the experiments. In particularly, thank to U. Stegmaier, P. Severloh, J. Stuckert and M. Steinbrück from KIT for the sample preparation, the preparation of the INRRO furnace and the scientific discussions, respectively. Thanks also to A. Kaestner, G. Frey, G. Kühne and S. Hartmann from PSI and E. Calcada from TU Munich for their help during the neutron radiography experiments.

REFERENCES

1   M. Steinbrück, J. Nucl. Mater. 334, 58 (2004).

2   M. Grosse, M. Steinbrueck, E. Lehmann, and P. Vontobel, Oxid. Metals 70, 149 (2008).

3   J. Stuckert, J. Birchley, M. Grosse, T. Haste, L. Sepold, and M. Steinbrueck M., Annals of Nuclear Energy 36, 183 (2009).

4   J. Stuckert, M. Grosse, L. Sepold, and M. Steinbrueck M., (2009), "Experimental results of reflood bundle test Quench-14 with M5® cladding tubes", Proceedings of the 17th International Conference on Nuclear Engineering (ICONE-17), Bruxelles, B, July 12-16, ICONE17-75266.

5   J. Stuckert, M. Grosse, L. Sepold, and M. Steinbrueck., (2009), "Experimental results of reflood bundle test QUENCH-15 with ZIRLOTM cladding tubes", Steinbrück, M., (Ed.), Proceedings of the 15th International QUENCH Workshop, Karlsruhe, November 3 - 5, Karlsruhe Institute of Technology, ISBN 978-3-923704-71-2.

H2 Storage and Hydrogen in Solids II

Mater. Res. Soc. Symp. Proc. Vol. 1262 © 2010 Materials Research Society   1262-W04-01

Nanostructured Metal Hydrides for Hydrogen Storage Studied by *In Situ* Synchrotron and Neutron Diffraction

V. A. Yartys[1,2], R.V. Denys[1], J.P. Maehlen[1], C.J. Webb[3], E. MacA. Gray[3], T. Blach[3],
A.A. Poletaev[1,2], J.K. Solberg[2], and O. Isnard[4]

[1]Institute for Energy Technology, P.O.Box 40, Kjeller, NO-2027, NORWAY
[2]Norwegian University of Science and Technology, Trondheim, NO-7491, NORWAY
[3]Queensland Micro- and Nanotechnology Centre, Griffith University, Nathan 4111, AUSTRALIA
[4]Institute Néel, CNRS/UJF, 38042 Grenoble, FRANCE

ABSTRACT

This work was focused on studies of the metal hydride materials having a potential in building hydrogen storage systems with high gravimetric and volumetric efficiencies of H storage and formed / decomposed with high rates of hydrogen exchange. *In situ* diffraction studies of the metal-hydrogen systems were explored as a valuable tool in probing both the mechanism of the phase-structural transformations and their kinetics. Two complementary techniques, namely Neutron Powder Diffraction (NPD) and Synchrotron X-ray diffraction (SR XRD) were utilised. High pressure *in situ* NPD studies were performed at D_2 pressures reaching 1000 bar at the D1B diffractometer accommodated at Institute Laue Langevin, Grenoble. The data of the time resolved *in situ* SR XRD were collected at the Swiss Norwegian Beam Lines, ESRF, Grenoble in the pressure range up to 50 bar H_2 at temperatures 20-400°C.

The systems studied by NPD at high pressures included deuterated Al-modified Laves-type C15 $ZrFe_{2-x}Al_x$ intermetallics with x = 0.02; 0.04 and 0.20 and the $CeNi_5-D_2$ system. D content, hysteresis of H uptake and release, unit cell expansion and stability of the hydrides systematically change with Al content. Deuteration exhibited a very fast kinetics; it resulted in increase of the unit cells volumes reaching 23.5 % for $ZrFe_{1.98}Al_{0.02}D_{2.9(1)}$ and associated with exclusive occupancy of the $Zr_2(Fe,Al)_2$ tetrahedra.

For $CeNi_5$ deuteration yielded a hexahydride $CeNi_5D_{6.2}$ (20°C, 776 bar D_2) and was accompanied by a nearly isotropic volume expansion reaching 30.1% ($\Delta a/a=10.0\%$; $\Delta c/c=7.5\%$). Deuterium atoms fill three different interstitial sites including Ce_2Ni_2, Ce_2Ni_3 and Ni_4. Significant hysteresis was observed on the first absorption-desorption cycle. This hysteresis decreased on the absorption-desorption cycling.

A different approach to the development of H storage systems is based on the hydrides of light elements, first of all the Mg-based ones. These systems were studied by SR XRD. Reactive ball milling in hydrogen (HRBM) allowed synthesis of the nanostructured Mg-based hydrides. The experimental parameters (P_{H2}, T, energy of milling, ball / sample ratio and balls size), significantly influence rate of hydrogenation. The studies confirmed (a) a completeness of hydrogenation of Mg into MgH_2; (b) indicated a partial transformation of the originally formed α-MgH_2 into a metastable γ-MgH_2 (a ratio α/γ was 3/1); (c) yielded the crystallite size for the main hydrogenation product, α-MgH_2, as close to 10 nm. Influence of the additives to Mg on the structure and hydrogen absorption/desorption properties and cycle behaviour of the composites was established and will be discussed in the paper.

INTRODUCTION

Recent R&D on hydrogen storage have resulted in a new method, "hybrid" H storage, yielding improved by up to 50 % overall H storage system efficiency [1]. In present work we have focused on metal hydrogen systems where one can significantly increase hydrogen storage capacity of the MH on application of high H_2 pressures. H storage capacities of the MH suitable for such systems are highly pressure-dependent, when pressures increase to a few hundred bar.

Characterisation of such systems can be efficiently performed by performing *in situ* diffraction studies of the metal-hydrogen interactions in the pressure cells loaded by hydrogen gas. Using deuterated samples and neutron scattering allows to "see" hydrogen environment and to probe metal-H/D and H-H/D-D interactions via establishing a pressure-dependant changes of the phase-structural composition of the materials. Such works, despite providing a very interesting and important scientific insight, meet obvious technical challenges. Thus, the list of particular metal hydride systems studied by *in situ* NPD at high D_2 pressures is rather limited. The system mostly focused during the earlier research is the $LaNi_5$-D_2 system where systematic studies at different applied high pressure experimental conditions were performed (see [2] as example). However, broadening of the materials field would be much desirable allowing overseeing a broader picture of the behaviours of the high pressure hydrides. Such a task was aimed in present work where a structural analogue of La-containing intermetallic, $CeNi_5$–D_2 system was studied by in situ NPD at pressures up to 1000 bar.

Zr-based Laves phase compounds $ZrFe_2$ and $ZrCo_2$ can be hydrogenated only using high applied hydrogen pressures, up to 3000 bar, giving $ZrFe(Co)_2D_{-3}$ [3]. These deuterides were studied *ex situ* by NPD and their crystal structures were solved. Al substitution for Fe in $ZrFe_2$ significantly modifies the hydrogenation behaviours decreasing equilibrium pressures of hydrogen and increasing stability of the hydrides [4]. The range of the studied $ZrFe_{2-x}Al_x$ materials was limited to the alloys with a rather large content of Al; thus, the equilibrium pressures of hydrogen desorption were well below 100 bar [4]. In present work we were focused on the experimental study of the systems in which Al content was very low, below 0.1 at.Al/f.u. $Zr(Fe,Al)_2$, thus aiming to synthesize the hydrides where the stability was expected to be only slightly increased as compared to $ZrFe_2$-H_2 system.

The other focus of the in situ studies of the metal-hydrogen systems is in studies of the nanoscale materials, as they exhibit extremely high specific surface and form nanocomposites consisting of small crystallites of a metal hydride and nanoscale dopants, facilitating hydrogen exchange rates and lowering kinetic barriers for the phase-structural transformations. Synthesis of the nanomaterials with controlled content of doping catalysers was proved to be especially successful for the Mg based alloys where complementing high H storage capacity of Mg, 7.6 wt.% H, by increased rates of H charge and discharge will help in building practical systems for H storage based on this light metal. *In situ* synchrotron XRD of the phase-structural transformations in the systems of Mg alloys and hydrogen gives valuable input in establishing the mechanism of the transformations, their kinetics as related to the nanostructuring. Reactive Ball Milling in hydrogen is a valuable tool in synthesis of the nanostructured Mg-based hydrides. These hydrides were studied by *in situ* SR XRD in our previous works [5,6]. In present paper we would like to report some new experimental data in the area. The focus will be on the V-catalysed processes of hydrogenation of Mg and its alloys.

EXPERIMENTAL DETAILS

Materials used: The $CeNi_5$ and $Zr(Fe,Al)_2$ alloys were prepared by arc melting using starting materials with a purity of at least 99.9%. X-ray diffraction pattern (Siemens D 5000 diffractometer, Cu $K\alpha_1$ radiation) showed that the samples prepared were single phases with correspondingly $CaCu_5$-type and $MgCu_2$ types of structures. The magnesium metal used was a grit of 50–150 mesh (0.1–0.3 mm) from Fluka (purity 99+%).

RBM in hydrogen: Hydrides were synthesized and processed by application of high-energy RBM in a Fritsch Pulverisette P6 planetary mill. The samples were placed into a custom-built Duplex SS 2377 (austenitic–ferritic steel) vial. The sample-to-balls weight ratio was ~ 1:80. The vial was initially evacuated and filled with 30 bar H_2. The milling was performed at 500 rpm. To monitor the hydrogenation process, the milling was paused (typically in 15-20 min intervals) and the vial, after cooling to room temperature, was connected to a Sieverts-type apparatus to monitor the amount of hydrogen absorbed; the vial was then refilled with hydrogen, and milling was continued. A complete hydrogenation of the Mg powder was reached in 6 hours, and in less than 2 h for its composite with 10-35 wt.% of the BCC-type $V_{75}Ti_{10}Zr_{7.5}Cr_{7.5}$ alloy.

***In situ* SR XRD experiments**: *In situ* SR-XRD studies of the Mg-based hydride nanocomposites were performed at the Swiss Norwegian Beam Lines (SNBL), at the European Synchrotron Radiation Facility (ESRF), Grenoble, France.
The measurements were performed using a set-up designed for *in situ* studies of the chemical processes occurring in hydrogen gas or in vacuum (see Ref. [7] and references therein). A small amount of the sample was put into a 0.5 mm quartz glass capillary (0.01 mm wall thickness), filling approximately 3-5 mm of the capillary. The capillary was seal-proof connected to the gaseous system using either a carbon ferrule, or a 1/8" Swagelok VCR fitting via a T-piece; the latter was attached to the goniometer head. Vacuum was created using a turbomolecular vacuum pump. Experimental data at BM01A were collected using a MAR2300 image plate detector. The wavelength and sample-to-detector distance were calibrated using LaB_6 as a reference material. The collection of one diffraction data set took 10 s. Powder diffraction data were analysed by the Rietveld whole-profile refinement method using the General Structure Analysis System (GSAS) [8] and Fullprof [9] softwares.

Crystallite size and strains: The sizes of the crystallites and strains in the materials were evaluated from the refinements of the Rietveld powder profile parameters using the following equations [10]: $D_V = \lambda/(\beta_s - cos\theta)$; $e = \beta_D/(4 - tan\ \theta)$, where D_V is a volume-weighted crystallite size, λ is the wavelength, β is the integral width of the reflection, θ is the Bragg angle and e is the microstrain. To obtain the integral width of the size-broadened and strain-broadened profiles, the size-broadening and micro-strain broadening profile coefficients were converted according to the procedure described in Ref. [10]. The instrumental contribution to the line broadening was evaluated by refining the profile parameters for LaB_6. During the refinements, only Lorentzian size-broadening and strain broadening profile parameters were refined.

In situ NPD measurements: Powder neutron diffraction data were collected with the D1B diffractometer at ILL, Grenoble ($\lambda = 2.52$ Å). More information about the in situ neutron powder diffraction can be found in the following review [11]. The scheme of the experimental setup is presented in Figure 1 and consists of a high-pressure Sievert's hydrogenator connected to a high-pressure sample cell made of a zero coherent scattering alloy (Zr-Ti) with a stainless steel inner liner. High pressures were achieved by the use of multistage, heat-based deuterium intensifier (maximum H_2 pressure 2 kbar). The *in situ* apparatus allowed simultaneous measurement of deuterium concentration in the sample and assessment of the crystal structure via neutron diffraction.

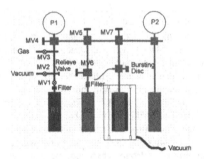

Figure 1. High pressure 1000 bar setup for the NPD studies. MV1 – A valve that isolates high purity hydrogen from the metal hydride reservoir (R1) to the LP section of the rig. MV2 – A valve that allows gas evacuation from the rig. MV3 – External gas supply valve. MV4 – A valve that isolates the LP and the HP sections of the rig. MV5 – A valve that isolates the HP reservoir from the rig. MV6 – A valve attached to the HP reservoir. MV7 – A valve isolating the sample volume (V2) from the reference volume (V1). MV8 – A valve isolating the sample volume from the rest of the rig. R1 – A volume filled with LaNi$_5$, used to store H$_2$ or D$_2$. The volume is used to purify the hydrogen and to intensify the gas to up to 200 bar using a heat pump. R2 – A volume filled by the alloy that stores H$_2$ at higher pressure (200 bar) and intensifies the pressure to 1000 bar using heat. V1 – Reference volume, temperature stabilized by water loop. V2 – Sample volume designed to operate inside a vacuum furnace.

RESULTS

CeNi$_5$ - D$_2$ system

Initial sample was characterised by SR XRD (SNBL; BM01A; λ=0.7350 Å) as a single phase alloy containing the CeNi$_5$ intermetallic compound with hexagonal unit cell; a=4.88415(7); c=4.00016(7) Å. First deuteration was tried without preliminary activation of the alloy and took place on application of high deuterium pressures. Our observations of the absorption-desorption processes can be summarized as follows:

- Both deuterium absorption and desorption are characterized by presence of clear plateaux on the *P-C* diagrams;
- A solid solution of deuterium in CeNi$_5$, α-CeNi$_5$D$_{-0.5}$, is formed first and extends until the level of pressure of 600 bar D$_2$. Then the β-CeNi$_5$D$_{-6}$ is formed.
- During the first deuteration, the midpoint of the plateau for the $\alpha \rightarrow \beta$ transformation at 20 °C is observed at 632 bar D$_2$;
- For the desorption, the midpoint of the plateau pressure for the $\beta \rightarrow \alpha$ transformation is significantly lowered, being 117 bar D$_2$ at 60 °C;

- During the second deuteration, midplateau pressure for $\alpha \to \beta$ transformation at 30 °C is significantly decreased to just 237 bar D_2;
- The observed regularities illustrating a significant hysteresis in D absorption/desorption concluded from analysis of the NPD pattern (see Figures 2, 3) are in good agreement with the reference data [12].

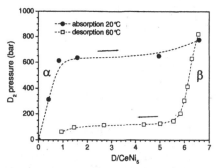

Figure 2. First deuterium absorption and desorption by $CeNi_5$ from the PND data.

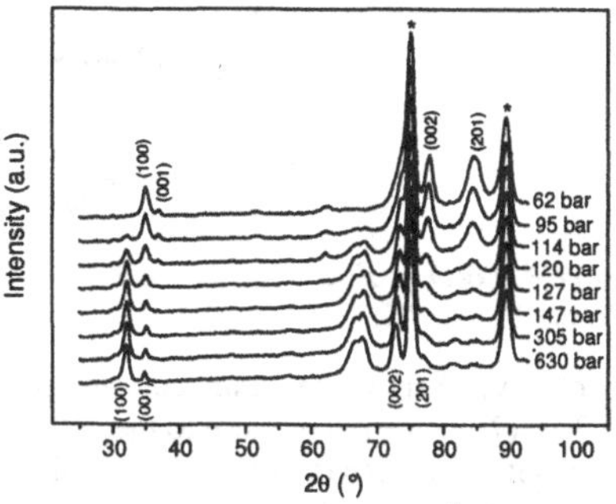

Figure 3. *In situ* NPD ($\lambda = 2.52$ Å) of deuterium desorption from the β-$CeNi_5D_{6.3}$ deuteride. α-$CeNi_5D_{\sim0.5}$ becomes visible at 147 bar D_2.

α-$CeNi_5D_{0.5}$: During the formation of the α-solid solution the hexagonal unit cell slightly expands; $\Delta a/a = 0.3\%$; $\Delta c/c = 0.6\%$; $\Delta V/V = 1.1\%$; $a = 4.8969(6)$; $c = 4.0261(6)$ Å; $V = 83.61(2)$ Å^3 [D1B; $\lambda = 2.52$ Å]. From Rietveld profile refinements of the NPD data we conclude that D atoms exclusively occupy the Ce_2Ni_2 $6m$ tetrahedra. Positional parameters for D are: $x = 0.118(5)$; $y = 2x$; $z = \frac{1}{2}$; $n = 0.085(7)$. Interestingly, only this interstitial site meets geometrical criterion for size limitation of the H-occupied sites, radius >0.4 Å: $r = 0.42$ Å. The other, vacant, sites have radii below 0.3 Å. It is worth to point out a difference between site occupancies in the α-$LaNi_5D_{0.42}$ [13] and α-$CeNi_5D_{0.5}$ solid solutions. Indeed, in the $LaNi_5$-based solid solution D atoms occupy two different sites, $6m$ (~2%) and $12n$ (3%), in contrast to the α-$CeNi_5D_{0.5}$ where only the $6m$ sites are partially filled by D atoms. From volumetric data of D absorption the limiting D content in α-solid solution is higher, 0.84 at.D/$CeNi_5$. An observed difference between the volumetric data and crystal structure refinements can be caused by presence of very small amount of the β-deuteride phase in the sample (ca. 5%) which starts to be formed already at 613 bar D_2. Such small quantities are difficult to confidently identify using the NPD data with a relatively low resolution.

β-$CeNi_5D_{6.2}$: Analysis of the NPD pattern showed a completeness of the transformation into the β-deuteride at pressure of 776 bar D_2. During the deuteration, the unit cell expands by 30.1 %, with $\Delta a/a = 10.0$ % and $\Delta c/c = 7.5$ %. Refinements yielded an excellent fit between the experimental and calculated data (see Figure 4). Three different interstitial sites are filled by D atoms, including Ce_2Ni2_2 tetrahedra $6m$; $Ce_2Ni1_2Ni_2$ sites $6i$ (half of the octahedra $Ce_2Ni1_2Ni2_2$) and tetrahedra Ni_4 $4h$. Crystal structure data for the $CeNi_5D_{6.3}$ are presented in Table I.

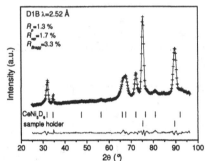

Figure 4. NPD profile refinement of the diffraction pattern for a saturated $CeNi_5D_{6.3}$ deuteride (776 bar D_2, 20 °C).

We observe that in the saturated deuteride $CeNi_5D_{6.3}$, the unit cell parameters, a=5.374(2); c=4.3013(7) Å, are rather close to the values for $LaNi_5D_6$ [14]. This contrasts to a large shrinking of the unit cells when to replace La in $LaNi_5$ by Ce to form $CeNi_5$. Thus, a Ce valence evolution towards a trivalent state is taking place upon hydrogenation to form $CeNi_5D_{-6}$. Furthermore, it seems that only *in situ* technique applied in present work was able to collect the correct crystallographic data for the deuteride. Indeed, in the work [12] where a stabilization of the high pressure hydride by SO_2 was applied, the experimentally measured values of the unit cell parameters were much smaller (a=5.368; c=4.235 Å) indicating a partial H desorption from the sample following a pressure decrease. Finally, we have observed appearance of anisotropic diffraction peak broadening upon first hydrogenation of $CeNi_5$, during α-$CeNi_5D_{-0.5}$ → β-$CeNi_5D_{-6}$ transformation. The anisotropic strain broadening is absent in the initial period of hydrogenation, until the β-deuteride appears and remains after further desorption-absorption cycling. Similar effect was observed in hydrogen-cycled pure and substituted $LaNi_5$ and attributed to dislocation-induced strain occurring after first hydrogenation [15]. The anisotropic peak broadening was modelled using the method applied in [16].

Table I.

Crystal structure data for β-$CeNi_5D_{6.3}$ (20°C, 776 bar D_2, D1B diffractometer at ILL).
Space group *P6/mmm*. a=5.374(2); c=4.3013(7) Å; V=107.58(5) Å^3.
Isotropic atomic displacement factors B_{iso} were fixed for Ce and Ni (1.0 Å^2) and for D (2.0 Å^2).

Atom	Site	x/a	y/b	z/c	Occupancy at. D/unit cell	Interatomic Me-D distances, Å	
Ce	1a	0	0	0			
Ni1	2b	⅓	⅔	0			
Ni2	3g	½	0	½			
D1	6i	½	0	0.143(3)	2.97(11) occup. 50%	Ce - D1 Ni1- D1 Ni2 - D1	2.76 1.67 1.54
D2	6m	0.137(2)	0.274(4)	½	2.66(8) occup. 44%	Ce - D2 Ni2 - D2	2.50 1.71
D3	4h	⅓	⅔	0.38(-)	0.62(4) occup. 16%	Ni1 - D3 Ni2 - D3	1.64 1.64

ZrFe$_{2-x}$Al$_x$ – D$_2$ system

ZrFe$_2$ intermetallic compound crystallizes with MgCu$_2$-type C15 FCC structure. Three different alloys containing 0.02; 0.04 and 0.20 at.Al/f.u. ZrFe$_{2-x}$Al$_x$ were synthesized by arc melting. SR XRD studies showed that Fe substitution by Al in Zr(Fe,Al)$_2$ alloys increases the unit cell parameter without change of the Laves-type of the crystal structure, from a=7.072 Å for ZrFe$_2$[4] to a=7.08492(8) Å for ZrFe$_{1.98}$Al$_{0.02}$, a=7.08658(7) Å for ZrFe$_{1.96}$Al$_{0.04}$ and a=7.0978(1) Å for ZrFe$_{1.80}$Al$_{0.20}$.

The hydrogenation/deuteration performance was studied at pressures of deuterium up to 1600 bar D$_2$. Corresponding deuterides were synthesized and studies by *in situ* neutron powder diffraction measurements at D1B instrument, ILL, Grenoble, France. From volumetric measurements performed in *in situ* setup, maximum D storage capacity reached 3.4 at.D/f.u. for ZrFe$_{1.98}$Al$_{0.02}$.

Deuteration resulted in expansion of the unit cells reaching $\Delta V/V$=23.5 % (a=7.6003(5) Å) for ZrFe$_{1.98}$Al$_{0.02}$D$_{2.9(1)}$; 23.2% (a=7.5966(5) Å) for ZrFe$_{1.96}$Al$_{0.04}$D$_{2.9(2)}$ and 18.6% for ZrFe$_{1.8}$Al$_{0.2}$D$_{2.7(2)}$ (a=7.5141(5) Å). Detailed data on the crystal structures are given in the Table II. Deuteration proceeds via formation of a two-phase region instead of a continuous solid solution and reaches saturation around 1000 bar D$_2$. A significant hysteresis was observed between the pressures of D$_2$ desorption and absorption during measurements of the isotherms (see Figure 5). From refinements of the NPD data (Figure 6) we conclude that D atoms occupy the Zr$_2$Fe(Al)$_2$ tetrahedra 96g (x=0.32; y=x; z=0.13; occupancy ~24%). Interatomic distances Me-D within the tetrahedra are: δ(Zr-D)=2.01-2.08; δ(Fe-D)=1.75 Å. The latter distances do not exclude a possibility of occupancy of the Al-substituted sites by H/D.

In situ studies demonstrated a very fast kinetics of H/D exchange in the material. From NPD data it was found that initial intermetallic alloys and their corresponding deuterides are ferromagnetic. Magnetic moments of Fe at room temperature slightly increase from the alloy (1.9 μ_B) to the corresponding deuteride (2.2 μ_B), similarly to ZrFe$_2$ based deuteride.

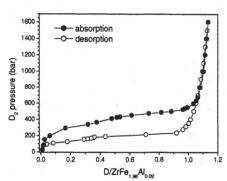

Figure 5. Deuterium absorption-desorption isotherms for ZrFe$_{1.98}$Al$_{0.02}$ at 20 °C.

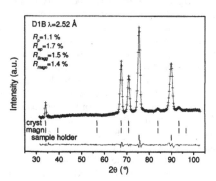

Figure 6. NPD profile refinements of the diffraction pattern for a saturated ZrFe$_{1.98}$Al$_{0.02}$D$_{2.9}$ deuteride (963 bar D$_2$, 20 °C).

Table II.

Crystal structure data for the $ZrFe_{2-x}Al_x$ (x=0.02; 0.04; 0.20) compounds and their corresponding deuterides (at 20 °C).   Space group *Fd-3m*. Zr in 8a (⅛, ⅛, ⅛); Fe(Al) in 16d (½, ½, ½).

Composition and D_2 pressure	*a*, Å	Δ*a*/*a*, %	Δ*V*/*V*, %	Δ*V*/at.D, Å³	Positions of D atoms (96g: *x, x, z*)	Interatomic Me-D distances, Å	
$ZrFe_{1.98}Al_{0.02}$	7.08492(8)						
$ZrFe_{1.98}Al_{0.02}D_{2.9(1)}$ 963 bar	7.6003(4)	7.3	23.5	3.59	x=0.319(1) z=0.126(1) occup. 24(1)%	Zr-D 2.08 Zr-D 2.01 Fe(Al)-D 1.75	
$ZrFe_{1.96}Al_{0.04}$	7.08658(7)						
$ZrFe_{1.96}Al_{0.04}D_{2.9(2)}$ 989 bar	7.5966(5)	7.2	23.2	3.56	x=0.319(1) z=0.126(1) occup. 24(1)%	Zr-D 2.09 Zr-D 2.00 Fe(Al)-D 1.75	
$ZrFe_{1.8}Al_{0.2}$	7.0978(1)						
$ZrFe_{1.8}Al_{0.2}D_{2.7(2)}$ 953 bar	7.5141(5)	5.9	18.6	3.09	x=0.324(7) z=0.129(4) occup. 23(2)%	Zr-D 2.11 Zr-D 1.98 Fe(Al)-D 1.70	

In situ SR XRD studies of hydrogen absorption-desorption processes in Mg-based hydride nanocomposites

Mg and Mg alloys can be easily hydrogenated during reactive ball milling (RBM) in H_2 gas, to yield nanocrystalline materials exhibiting excellent H sorption performances [6, 17]. The hydrogenation and dehydrogenation of the materials obtained by HRBM of pure Mg and its nanocomposites with a V-based BCC alloy, $V_{75}Ti_{10}Zr_{7.5}Cr_{7.5}$ (for simplicity, it further referred to in the paper as a V alloy), were experimentally studied by time resolved *in situ* synchrotron X-ray diffraction (SR XRD), in a temperature range from 20 to 400°C and at H_2 pressures up to 15 bar. In particular, the mechanism, the kinetics of the phase-structural transformations and the microstructural evolution of the samples during hydrogen uptake and release were characterized.

The main constituents of the hydrogenated by application of the RBM in hydrogen Mg and Mg-V alloy composites are two allotropic modifications of magnesium dihydride, the rutile-type α-MgH_2 and the high-pressure γ-MgH_2 (α-PbO_2 type) in ratio (appr. 3:1), and CaF_2-type vanadium dihydride (in Mg-V nanocomposites). The crystallite sizes of magnesium dihydrides were evaluated from the XRD peak broadening to be appr. 7 nm.

During the *in situ* SR XRD experiments the mechanochemically hydrogenated Mg and Mg-V alloy were subjected to absorption-desorption cycling at temperatures 25-300 °C and pressures 0-15 bar H_2. Evolution of SR XRD pattern of nanocrystalline MgH_2 as a function of temperature during the hydrogen desorption is shown in Fig.7. Fig.8 summarizes changes in phase abundances and allows identifying three contributing to the mechanism of decomposition processes: (1) a metastable hydride γ-MgH$_2$ partially transforms into α-MgH$_2$; (2) in parallel, it also directly decomposes to Mg and H$_2$; (3) α-MgH$_2$ directly decomposes to Mg. These three events were observed during the hydrogen desorption from the V-containing nanocomposites. However, the decomposition temperatures appear to be substantially, by $\sim$100°C, lower than for the individual nanocrystalline MgH$_2$. In addition to the MgH$_2 \rightarrow$ Mg transformation, a step-by-step transformation γ-VH$_2$ (FCC) $\rightarrow \beta_2$-VH$_{0.7}$ (BCC) $\rightarrow \beta_1$-VH$_{0.5}$ (BCT) $\rightarrow$ V (BCC) has been evidenced.

Peak broadening analysis of *in situ* SR XRD data showed a continuous increase in the crystallite sizes of MgH$_2$ hydrides during their heating in vacuum, from $\sim$7 nm at room temperature up to $\sim$20 nm at 350 °C in the end of decomposition process. The crystallite sizes of Mg metal formed during the thermal desorption from the RBM MgH$_2$ exceed 150 nm. On further re-hydrogenation, large crystallites of α-MgH$_2$, with sizes above 100 nm, are formed.

On subsequent isothermal re-hydrogenation/dehydrogenation cycling, the RBM materials show enhanced kinetics of the transformations. For Mg metal, its re-hydrogenation at 300 °C starts from an instant formation of the MgH$_x$ solid solution as the first stage of the H uptake process. The following hydride formation proceeds in accordance with the mechanism of nucleation and growth of α-MgH$_2$ occurring in the bulk of the Mg particles via a grain boundary attack.

Presence of V-based alloy in the nanocomposite leads to dramatic improvement of hydrogenation kinetics of Mg and reduction of hydrogenation temperature from 300°C to room temperature [16]. Such improvement of hydrogenation kinetics is caused by catalytic and exothermic heat effects of hydrogen absorption by the V-based alloy.

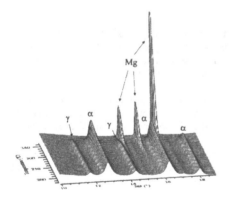

Figure 7. *In situ* SR XRD pattern (λ=0.7114(1) Å) of hydrogen vacuum thermal desorption from the RBM MgH$_2$.

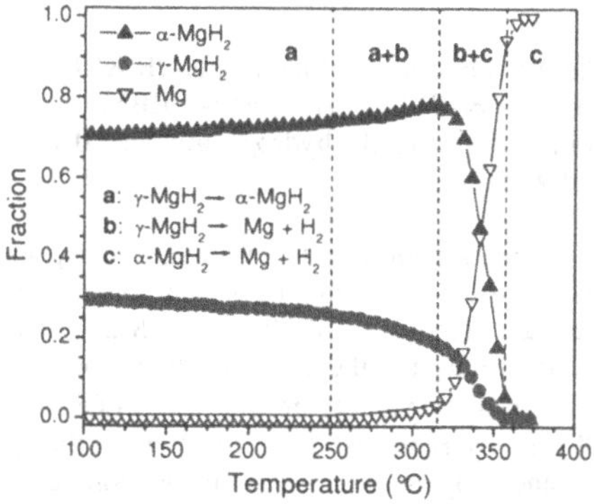

Figure 8. Changes in phase composition of the RBM MgH$_2$ during hydrogen thermal desorption from Rietveld refinements of *in situ* SR-XRD data.

Present study clearly shows that synthesis of the nanostructured composites of Mg and V-based alloys dramatically facilitates the rates of H uptake by the Mg-based materials and significantly decreases the reaction temperatures; in all cases 50–80% hydrogenation of Mg was observed already at room temperature. This facilitation has been achieved as a result of significant decrease in the particle size during the HRBM, thus decreasing diffusion lengths for hydrogen in the material, and, also by achieving synergy in hydrogenation of V and Mg. Hydrogen absorption by the Mg-V composites starts from instant formation of the V-based hydrides, which promote the hydrogenation of Mg. As can be seen from Figures 8 and 9, at the first stages of the hydrogenation process, an instant formation of the β_1-$VH_{0.5}$ is observed; it is followed by a fast transformation into the β_2-$VH_{0.7}$. Hydrogenation of Mg leading to the formation of α-MgH_2 has slower rate and proceeds after the vanadium monohydride is formed. The rate of the transformation $V \rightarrow \beta_2$-$VH_{0.7}$ was found to be two orders of magnitude higher than that for the hydrogenation of Mg.

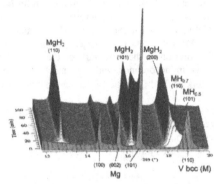

Figure 9. Evolution of the in situ SR XRD pattern (λ=0.7114(1) Å) of Mg-V RBM composite during the hydrogenation at 80°C and 9 bar H_2.

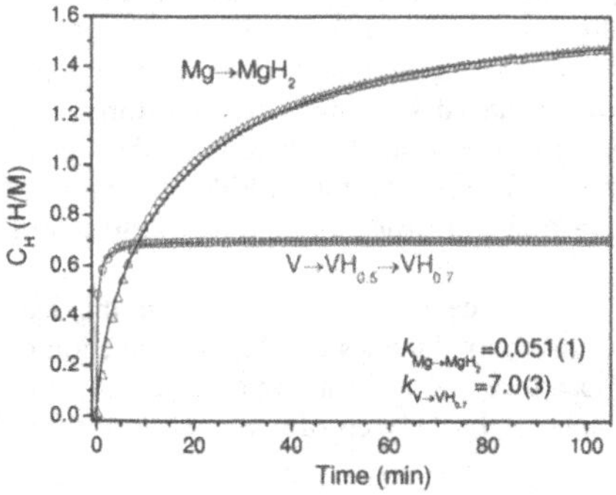

Figure 10. Hydrogenation kinetics of the individual components of the Mg-V RBM composite from the Rietveld refinements of the *in situ* SR XRD data.

The catalytic effect of V on the hydrogenation of Mg can be identified as follows: (1) V acts as a source of atomic H, because of an easy H_2 dissociation on its surface and (2) formation of extra centers for the nucleation of MgH_2 on an interface between the Mg and V alloy, where V acts as a catalyst of the hydrogenation process.

On hydrogen desorption, the sequence of events is opposite; α-MgH_2 decomposes first followed by H release from the $VH_{-0.5}$. Substantial decrease in desorption temperature can be explained by a facilitated nucleation of Mg phase in presence of V nanoparticles.

CONCLUSIONS

In conclusion, this work clearly demonstrates a great potential of the *in situ* diffraction techniques, neutron powder diffraction and synchrotron X-ray powder diffraction, in probing the mechanism of hydrogenation to form solid hydrogen storage materials. Neutron scattering at high pressures not only provides an insight into the metal-hydrogen bonding by establishing hydrogen neighborhood and relative population of different sites, but also shows a thermodynamic difference between the hydrogenation and hydrogen desorption processes and changes in the behaviors of cycling of hydrogen uptake and release. When complemented by *in situ* SR XRD, extra features of the metal-hydrogen interactions become uncovered, by exploring benefits from significantly improved time resolution of the processes, high sensitivity in determining formation of the phase constituents and excellent accuracy in yielding the crystallographic data. Thus, further development of combined *in situ* NPD and SR XRD studies of the metal-hydrogen systems should be considered to be a very relevant and important task in discovery and optimization of the novel and advanced hydrogen storage materials.

ACKNOWLEDGEMENTS

We are grateful to the staff of Swiss-Norwegian Beam Lines, ESRF, Grenoble for skillful assistance during the synchrotron X-Ray diffraction experiments.

REFERENCES

1. N. Takeichi, H. Senoh, T. Yokota, H. Tsuruta, K. Hamada, H. T. Takeshita, H.Tanaka, T. Kiyobayashi, T. Takano, N. Kuriyama. *International Journal of Hydrogen Energy* **28**, 1121 (2003).
2. E. MacA. Gray, E.H. Kisi, R.I. Smith. *J. Alloys and Compounds* **293–295**, 135 (1999).
3. V. Paul-Boncour, F. Bourée-Vigneron, S.M. Filipek, I. Marchuk, I. Jacob, A. Percheron-Guégan. *Journal of Alloys and Compounds* **356–357**, 69 (2003).
4. M. Bereznitsky, I. Jacob, J. Bloch, M.H. Mintz. *J. Alloys and Compounds* **351**, 180 (2003).
5. R.V. Denys, A.B. Riabov, J.P. Maehlen, M.V. Lototsky, J.K. Solberg, V.A. Yartys. *Acta Materialia* **57**, 3989 (2009).
6. R.V. Denys, A.A. Poletaev, J.K. Solberg, B.P. Tarasov, V.A. Yartys. *Acta Materialia* **58**, 2510 (2010).
7. J.P. Maehlen, V.A. Yartys, R.V. Denys, M. Fichtner, Ch. Frommen, B.M. Bulychev, P. Pattison, H. Emerich, Y.E. Filinchuk, D. Chernyshov. *J. Alloys and Compounds* **446–447**, 280 (2007).
8. A.C. Larson, R.B. von Dreele. General structure analysis system (GSAS), LANSCE, MS-H, **805**, 1994.
9. J. Rodriguez-Carvajal. XV Congress of International Union of Crystallography, Satellite Meeting on Powder Diffraction Toulouse, France, 1990, p. 127.
10. D. Balzar, N. Audebrand, M.R. Daymond, A. Fitch, A. Hewat, J.I. Langford, A. Le Bail, D. Louër, O. Masson, C.N. McCowan, N.C. Popa, P.W. Stephens and B.H. Toby. *J. Appl. Cryst.* **37**, 911 (2004).
11. O. Isnard, *C. Rend. Phys.* **8**, 789 (2007).
12. S.N. Klyamkin, N.S. Zakharkina. *J. Alloys and Compounds* **361**, 200 (2003).
13. M. Latroche, J.-M. Joubert, A. Percheron-Guegan, F. Bouree-Vigneron. *J. Solid State Chemistry* **177**, 1219 (2004).
14. M.P. Pitt, E. MacA. Gray, E.H. Kisi, B.A. Hunter. *J. Alloys and Compounds* **293–295**, 118 (1999).
15. R. Cerny, J.-M. Joubert, M. Latroche, A. Percheron-Guegan, K. Yvon. *J. Appl. Crystallogr.* **33**, 997 (2000).
16. M. Latroche, J. Rodriguez-Carvajal, A. Percheron-Guegan, F. Bouree-Vigneron. *J. Alloys and Compounds* **218**, 64 (1995).
17. M.V. Lototsky, R.V. Denys, V.A. Yartys. *Int. J. Energy Res.* **33(13)**, 1114 (2009).

Mater. Res. Soc. Symp. Proc. Vol. 1262 © 2010 Materials Research Society   1262-W04-02

Magnetic state in iron hydride under pressure studied by X-ray magnetic circular dichroism at the Fe K-edge

N. Ishimatsu[1], Y. Matsushima[1], H. Maruyama[1], T. Tsumuraya[2,*], T. Oguchi[2], N. Kawamura[3], M. Mizumaki[3], T. Matsuoka[3], and K. Takemura[4]

[1]Gradate School of Science, Hiroshima University, 1-3-1 Kagamiyama, Higashi-Hiroshima 739-8526, Japan
[2]Gradate School of Advanced Science of Matter, Hiroshima University, 1-3-1 Kagamiyama, Higashi-Hiroshima 739-8530, Japan
[3]Japan Synchrotron Research Institute/SPring-8, 1-1-1, Kouto, Sayo, 679-5198, Japan
[4]National Institute for Materials Science, 1-1 Namiki, Tsukuba, 305-0044, Japan
*present address: Department of Physics and Astronomy, Northwestern University, Evanston, IL 60208-3112, USA

ABSTRACT

In order to study magnetic states in Fe-hydride under pressure, X-ray magnetic circular dichroism (XMCD) at the Fe K-edge has been measured up to 27.5 GPa. As a result, hydrogenation from bcc-Fe to dhcp-FeH occurs within a narrow region of 3.2-3.8 GPa, which is clearly observed by the dichroic profile in dhcp-FeH differing from that in bcc-Fe. Influence of H atoms on Fe 3d and 4p electronic states is discussed using the pressure-dependent XMCD and the first-principles calculation.

INTRODUCTION

Metal hydrogen system has recently attracted scientific and technological interests because of its increasing demand for hydrogen storage materials. Insertion of hydrogen atoms leads to volume expansion of the host metal and/or structural transition, which results in the drastic changes in brittleness, electric resistivity, optical transparency, and so on [1]. In the case of 3d ferromagnetic transition metals, Fe, Co and Ni, hydrogenation also gives rise to some changes in their magnetic properties [2,3]. Influence of the hydrogenation on the electronic and magnetic structure has an important role for these phenomena.

Hydrogen solubility of the 3d transition metals is significantly low at atmospheric pressures; however, the hydrogenation further proceeds under the order of GPa hydrogen pressures. In the case of iron, bcc-Fe is rapidly hydrogenated to ferromagnetic FeH in H_2 fluid under pressure higher than 3.5 GPa.[4] FeH takes a double hcp (dhcp) structure (space group: $P6_3/mmc$), in which Fe atoms occupy 2a and 2c sites whereas H atoms occupy 4f site [5]. It has been predicted that H atoms are slightly displaced towards Fe at 2a site from the center position [2,3]. Mössbauer spectroscopy has revealed two different hyperfine fields, indicating that Fe atoms carry different magnitude of magnetic moment [6]. These magnetic properties contrast with the non-magnetic state in hcp Fe observed under pressures higher than 14 GPa [7]. To study the magnetic property in FeH, it is crucial to understand the modification of the electronic structure due to the hydrogenation and stability of the ferromagnetism under pressure.

For this purpose, X-ray magnetic circular dichroism (XMCD) is a powerful spectroscopic technique because it enables us to probe magnetically polarized electronic states with element and orbital selectivity. Recently, XMCD has been applied to studies under high pressure using the $3d$ transition metal K absorption edge located at hard X-ray energy range, typically $E > 6.5$ keV [7,8]. Therefore, we emphasize that this method is useful for the study on ferromagnetic $3d$ transition metal hydrides. The K-edge XMCD spectrum appears owning to the spin-polarization and spin-orbit coupling at $4p$ orbital. The K-edge XMCD is also sensitive to the spin-polarized $3d$ states because of hybridization between $4p$ and $3d$ orbitals. In this study, we measured XMCD of dhcp-FeH together with X-ray absorption near edge structure (XANES) at the Fe K-edge up to 27.5 GPa. We also performed first-principles calculation in order to investigate the electronic structure and the profile of XMCD. We discuss the electronic structure and magnetic state of FeH and the stability of the ferromagnetic phase under pressure.

EXPERIMENTAL PROCEDURE AND THEORETICAL CALCULATION

Diamond anvil cell (DAC) made of Cu-Be alloy was employed to apply high pressures. Polycrystalline bcc-Fe foil was used as an initial sample for hydrogenation. Tiny foil of about $50 \times 80 \times 5 \ \mu m^3$ in size was prepared and put into the DAC together with H_2 fluid which was initially pressurized to 0.18 GPa. For loading the H_2 fluid, we utilized a versatile gas-loading system [9]. The H_2 fluid worked as the source of the hydrogenation and as the pressure-transmitting medium. A pair of thin diamond anvils with 2.0 mm total thickness was adopted to reduce X-ray absorption due to the anvils. Culet diameter of the anvils was 450 μm. The pressure applied to the sample was evaluated by the conventional ruby fluorescence method [10].

XANES and XMCD experiment under high pressure was carried out on the beamline 39XU at the SPring-8 facility [11]. The incident X-ray beam was focused by Kirkpatrick and Baez mirror [12]. The beam size was $7 \times 7 \ \mu m^2$ at the sample position. The spectra were measured by using the helicity-modulation method [11]. A diamond phase retarder of 0.45 mm in thickness was employed to produce X-ray beams with a high degree of circular polarization, $P_c \gtrsim 0.9$ [13]. Magnetic field of 0.6 T was applied parallel to the incident X-ray beam and surface normal of the sample.

Theoretical XANES and XMCD spectra of dhcp-FeH were calculated by first-principles calculations based on the density functional theory with the all-electron full-potential linear augmented plane wave (FLAPW) method [14]. Exchange and correlation were treated within its spin-polarized form with the generalized gradient approximation (GGA). A uniform k-mesh set of $24 \times 24 \times 12$ was used for the integration in the Brillouin zone. Muffin-tin sphere radii were set to be 1.0Å for Fe and 0.8Å for H. Plane-wave cut-offs were 20 and 150 Ry for the LAPW basis functions and the potential and charge density, respectively. Before calculating theoretical spectra, we performed structural optimization for dhcp-FeH and obtained the lattice constants of $a = 2.651$ Å and $c = 8.67$ Å. The probability of transition per unit time was given by Fermi's golden rule and the transition matrix elements were calculated with the inclusion of both electric dipole and quadrupole transitions.

RESULTS AND DISCUSSION

Figure 1 shows the pressure dependence of XANES and XMCD at the Fe K-edge. The XANES profile of bcc-Fe has the shoulder structure A at 7.113 keV. XMCD of bcc-Fe gives a dispersion-type profile which mainly consists of the positive peak C and negative peak D. The XANES and XMCD of bcc-Fe remain unchanged up to 3.2 GPa. However, both spectra are

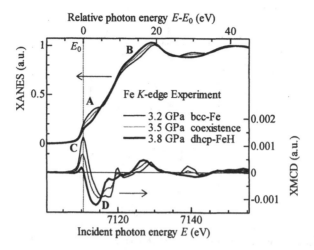

Figure 1. Pressure dependence of the Fe K-edge XANES and XMCD spectra. The vertical line indicates the absorption edge $E_0 = 7.1105$ keV.

abruptly modified at 3.5 GPa and then are completely transformed into those corresponding to dhcp-FeH at the pressures above 3.8 GPa. Therefore, the hydrogenation suddenly occurs and proceeds in the narrow pressure range less than 0.6 GPa. The intermediate XANES and XMCD spectra at 3.5 GPa are interpreted by the coexistence of the two phases. The coexistence has been confirmed by the X-ray powder diffraction experiment [5,15]. The magnitude of XMCD for dhcp-FeH is comparable to that for bcc-Fe, and hence dhcp-FeH is ferromagnetic. Compared with the spectral features of XANES for bcc-Fe, shoulder **A** of dhcp-FeH is rather small whereas the significant increase is observed at the crest **B** near the absorption maximum. As for the XMCD of dhcp-FeH, the spectrum is characterized by the sharp negative peak **D** together with the small positive peak **C** near the absorption edge $E_0 = 7.1105$ keV.

Calculated XANES and XMCD spectra of dhcp-FeH are displayed in Fig. 2. The calculations show that XMCD of dhcp-FeH provides different profile depending on the direction of the magnetization M. XMCD($M//c$) denotes the spectrum in the case of M parallel to the easy axis *i.e.* c-axis, and XMCD($M//a$) corresponds to M parallel to a-axis. The strong dependence on the direction of M is probably due to the magneto-crystalline anisotropy of the dhcp structure. Since the sample used is polycrystalline foil, we assume random distribution of M with respect to the dhcp crystal. The average of the calculated XMCD profile is evaluated by (XMCD($M//c$)+ 2×XMCD($M//a$)) /3 for comparison with the experiment. As shown in Fig. 2, the calculated XANES and XMCD fairly reproduce the experimental profiles. Although the calculation does not exhibit the small positive peak **C**, the sharp negative is in good agreement with the peak **D** in the experiment. This reproducibility indicates that the electronic structure and the magnetic states of dhcp-FeH are well explained by our theoretical calculation.

XANES and XMCD spectra of dhcp-FeH are interpreted by the modification of the electronic structure which is influenced by both structural transition and the hydrogenation. Firstly, we point out that the electronic structure based on the bcc structure is transformed into

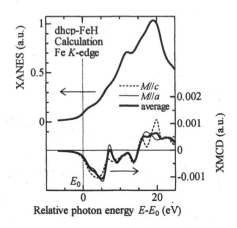

Figure 2. calculated Fe K-edge XANES and XMCD spectra of dhcp-FeH.

one based on the closed pack structure. The first-principles calculation demonstrates that the electronic structure of dhcp-FeH is mostly classified as strong ferromagnetism, where the $3d$ majority (minority) DOS is fully (partially) occupied. It clearly differs from the electronic structure of bcc-Fe classified as weak ferromagnetism. The electronic structure is responsible for the change of the XMCD spectrum from the dispersion-type profile to the profile mainly consisting of the negative peak **D**. Secondly, the electron filling of d-state is also important to elucidate the XMCD spectrum. The tight-binding calculation has reported that the electron filling of d-state causes the decrease in the positive peak **C** near E_0 of the K-edge XMCD [16]. For fcc $Fe_{1-x}Ni_x$ alloys, the similar behavior has been confirmed when the Ni content x increases [17]. Therefore, the small peak **C** indicates an increase in d-band occupancy of dhcp-FeH. We infer that the H atom provides the additional electrons. Our calculation demonstrates that there is 0.1 extra electron in the number of Fe $3d$ electrons compared with that of bcc-Fe.

In metal hydride systems, hydrogen atoms form bonding and anti-bonding states with host elements. In the case of the electronic structure of dhcp-FeH, the bonding state is located at about E_F −7 eV, and anti-bonding state is formed above E_F, where E_F is the Fermi energy. Both states mainly consist of H $1s$ and Fe $3d$ orbitals. Furthermore, our calculation shows that the anti-bonding state results in a remarkable increase in Fe $3d$ density of states (DOS) at the vicinity of E_F +6 eV. The energy range of the Fe $3d$ DOS is almost similar to the width of the negative XMCD peak **D** which appears in the narrow energy range up to E_0+7 eV. In addition, the K-edge XMCD is sensitive to the $3d$ electronic state via p-d hybridization so that the sharp peak **D** is probably interpreted as the newly-formed $3d$ state. For further discussion, it is valuable to separately investigate the magnetic states of Fe at $2a$ and $2c$ sites. However, the experimental XMCD is the average of two spectra originating from each site. The calculated spectra indicate that Fe at $2c$ site gives slightly larger amplitude of peak **D** than Fe at $2a$ site.

The shoulder structure **A** of the XANES is ascribed to the p-d hybridization with the neighboring Fe atoms [18]. As shown in Fig. 1, the shoulder **A** significantly decreases accompanied with the transition from bcc-Fe to dhcp-FeH. It is elucidated that the hydrogenation suppresses the p-d hybridization and reduces the unoccupied Fe $4p$ DOS near E_F. These phenomena also lead to the upward shift of Fe $4p$ DOS. The latter effect is recognized by the remarkable enhancement of the crest **B**, indicating that Fe $4p$ DOS increases at the energy range above E_F +10 eV. We consider that the change in Fe $4p$ DOS near E_F is also associated with the anti-bonding state.

In this study, pressure dependence of XMCD is measured up to the maximum pressure of 27.5 GPa. At the pressures above the hydrogenation $P>3.8$ GPa, the amplitude of XMCD gradually decreases with increasing pressure. At 27.5 GPa, the amplitude is about 16 % of that at 3.8 GPa. Extrapolating the pressure dependence, we determine the critical pressure where the amplitude of XMCD vanishes. The critical pressure is estimated to be 29.5 GPa. The disappearance of XMCD means that dhcp-FeH is not ferromagnetic. We consider that dhcp-FeH is in a paramagnetic state above the critical pressure, and the spin polarization is small enough to reduce the Curie temperature below room temperature.

CONCLUSIONS

We reported the magnetic state for dhcp-FeH and its pressure dependence up to 27.5 GPa, using XMCD at the Fe K-edge and the first-principles calculation. The onset of the hydrogenation to dhcp-FeH was observed within the narrow pressure region 3.2-3.8 GPa. This study has revealed the remarkable differences of XANES and XMCD spectra between bcc-Fe and dhcp-FeH. The XANES spectra of dhcp-FeH indicate that Fe $4p$ DOS near E_F decreases, and the sharp XMCD peak **D** is probably associated with Fe $3d$ DOS near E_F due to the H $1s$ anti-bonding state. The pressure-induced decrease in XMCD shows that the ferromagnetic state would disappear at 29.5 GPa. This study demonstrates that the XMCD is a useful technique to prove the electronic and magnetic states of metal hydrides even at high pressure.

ACKNOWLEDGMENTS

One of the authors (N. I.) appreciates Dr. Naohisa Hirao and Dr. Takaya Mitsui for their helpful discussion. This work was supported by the Grants of the NEDO project "Advanced Fundamental Research on Hydrogen Storage Materials." The XMCD measurement was performed at SPring-8 with the approval of PRC-JASRI (Nos. 2008B1284 and 2009B1011).

REFERENCES

1. Y. Fukai, "The Metal-Hydrogen System" 2nd ed. Springer, Berlin (2005).
2. V.E. Antonov, J. Alloys Comp., **330-332**, 110 (2002).
3. V.E. Antonov, M. Baier, B. Dorner, V.K. Fedotov, G. Grosse, A.I. Kolesnikov, E.G. Ponyatovsky, G. Schneider, and F.E. Wagner, J. Phys.: Condens. Matter, **14**, 6427 (2002).
4. V.E. Antonov, I.T. Belash, V.F. Degtyareva, E.G. Ponyatovskii, V.I. Shiryaev, Sov. Phys. Dokl. **25**, 490 (1980).
5. J.V. Badding, R.J. Hemley, H.K. Mao, Science, **253**, 421 (1991).
6. I. Choe, R. Ingalls, J.M. Brown, Y. Sato-Sorensen, R. Mills, Phys. Rev. B, **44**, 1 (1991).

7. N. Ishimatsu, H. Maruyama, N. Kawamura, M. Suzuki, Y. Ohishi, O. Shimomura, J. Phys. Soc. Jpn., **76**, 064703 (2007).
8. Y. Ding, D. Haskel, Y.C. Tseng, E. Kaneshita, M. van Veenendaal, J.F. Mitchell, S.V. Sinogeikin, V. Prakapenka, H.K. Mao, Phys. Rev. Lett. **102**, 237201 (2009).
9. K. Takemura, P.C. Sahu, Y. Kunii, Y. Toma, Rev. Sci. Instrum., **72**, 3873 (2001).
10. H. K. Mao, J. Xu, P. M. Bell, J. Geophys. Res. **91**, 4673 (1986).
11. N. Kawamura, N. Ishimatsu, H. Maruyama, J. Synchrotron Rad., **16**, 730 (2009).
12. H. Yumoto et al., Proc. SPIE 7448, 74480Z (2009).
13. K. Hirano, K. Izumi, T. Ishikawa, S. Annaka, S. Kikuta, Jpn. J. Appl. Phys. **30**, L407 (1991).
14. T. Tsumuraya, T. Shishidou and T. Oguchi, Phys. Rev. B **77**, 235114 (2008).
15. N. Hirao, T. Kondo, E. Ohtani, K. Takemura, T. Kikegawa, Geophys. Res. Lett. **31**, L06616 (2004).
16. J. Igarashi, K. Hirai, Phys. Rev. B **53**, 6442 (1996).
17. H. Sakurai, F. Itoh, H. Maruyama, A. Koizumi, K. Kobayashi, H. Yamazaki, Y. Tanji, H. Kawata, J. Phys. Soc. Jpn., **62**, 459 (1993).
18. J. Chaboy, J. Garcia, A. Marcelli, J. Magn. Magn. Mater., **166**, 149 (1997).

Mater. Res. Soc. Symp. Proc. Vol. 1262 © 2010 Materials Research Society   1262-W04-03

Formation of Methane Hydrates from Super-compressed Water and Methane Mixtures

Jing-Yin Chen[1] and Choong-Shik Yoo[1]
[1]Institute for Shock Physics and Department of Chemistry, Washington State University, Pullman, WA 99163, U.S.A.

ABSTRACT

Understanding the high-pressure kinetics associated with the formation of methane hydrates is critical to the practical use of the most abundant energy resource on earth. In this study, we have studied, for the first time, the compression rate dependence on the formation of methane hydrates under pressures, using *dynamic*-Diamond Anvil Cell (*d*-DAC) coupled with a high-speed microphotography and a confocal micro-Raman spectroscopy. The time-resolved optical images and Raman spectra indicate that the pressure-induced formation of methane hydrate depends on the compression rate and the peak pressure. At the compression rate of around 5 to 10 GPa/s, methane hydrate phase II (MH-II) forms from super-compressed water within the stability field of ice VI between 0.9 GPa and 2.0 GPa. This is due to a relatively slow rate of the hydrate formation below 0.9 GPa and a relatively fast rate of the water solidification above 2.0 GPa. The fact that methane hydrate forms from super-compressed water underscores a diffusion-controlled growth, which accelerates with pressure because of the enhanced miscibility between methane and super-compressed water.

INTRODUCTION

Natural gas hydrates are a new class of energy source, made of hydrogen-bonded cages of water molecules containing a wide range of guest molecules (CH_4, H_2, CO_2, *etc.*). Methane hydrates are the most abundant in nature and the largest energy resource of all fossil fuels. On the other hand, because methane is a greenhouse gas that is approximately 10 times more powerful than carbon dioxide, its release could potentially result in abrupt climate change and thereby living conditions on the Earth. Because of these reasons, there have been extensive studies on methane hydrates [1-8].

Methane hydrates are stable in ocean floor sediments at water depths greater than 300 meters (~ 3 MPa) and are found most at the water depth between 2000 and 3000 m (20-30 MPa). Because it is stable only within a small pressure-temperature domain, recovering and transporting the hydrate or methane in a controlled manner is a significant technical challenge. An uncontrolled sudden release of methane gas could cause geological and industrial hazards, as well as significant environmental consequences. Addressing this technical challenge requires fundamental understandings of hydrogen bonding and disorder, and, more importantly, chemical mechanisms and kinetics governing the formation and decomposition of methane hydrates under pressures.

Recent development of dynamic-diamond anvil cells (*d*-DAC)[9-11] enables us to study high-pressure kinetics associated with the crystal growth, phase transitions, and chemical reactions, over a wide range of compression rates and pressures. Therefore, using *d*-DAC coupled with the Raman spectroscopy and high-speed photography, we have studied the chemical mechanism and kinetics of hydrate formation and decomposition and the phase transitions and metastable structures. In this paper, we present the experimental evidence that the formation of methane hydrate is strongly dependent on the compression rate and the stability field of water.

EXPERIMENT

Highly pure methane gas (>99.999%, Advanced Specialty Gases) was used without future purification. Using a micro-syringe with a stainless steel needle, a drop of pure distilled water (< 10 µl) was loaded into a small (0.08-0.13 mm) hole of a pre-indented stainless steel gasket mounted between two opposed diamond anvils with 0.3 or 0.5 mm flats, together with a few small ruby chips for pressure measurements. A small bubble was then introduced in the sample chamber for the space of methane. Methane gas was then loaded into the sample chamber using a high-pressure gas loader developed at Washington State University.

In this paper, we describe four different experiments performed at the identical concentration. The same initial condition was prepared from the same sample, first by melting (or decomposing) methane hydrates at 75 °C to methane and water and then slowly cooling the mixture to ambient temperature. In this way, the sample is recovered back to the original methane-water mixture without methane hydrates. We used confocal micro-Raman spectroscopy to characterize chemical species, and phases, of ice, methane hydrates, and methane, as well as the size of water cages [1-6]. Raman spectra were obtained in a back scattering geometry with a 0.8 cm^{-1} spectral resolutions and using an Ar$^+$ ion laser (Spectra Physics) at the 514.5 nm excitation wavelength. The pressure of the sample was determined by measuring the R1 line of Ruby luminescence and using the quasi-hydrostatic ruby pressure scale. The formation of methane hydrates was recorded *in-situ*, using a high-speed camera (Photron FASTCAM APX RS) or a regular fast camera (AVT Marlin F-131B) depending on the rate of compression and transitions.

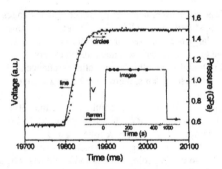

Figure 1. The applied waveform (solid lines) and the pressure measurements (solid circles) as a function of time. The inset shows the whole 1000 s waveform and event markers. The stars illustrate that the Raman are taken before and after during each run and the solid hexagons show the real timing of each microphotograph shown in Fig. 3.

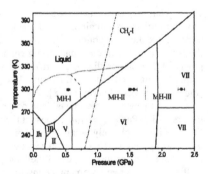

Figure 2. Phase diagrams of H$_2$O (solid lines), methane (dashed dotted lines) and methane hydrate (dotted lines)[1-4]. The solid circles represented the achieved maximum pressure during each run.

88

Our *d*-DAC incorporates three piezo-actuators (APC International, Ltd.) into a conventional DAC. The peak and modulating pressures and the compression rate of *d*-DAC are controlled by the shape and magnitude of the input electric signal to these actuators, as shown in Fig. 1. We used a function generator and a power amplifier to tailor the input signal of the piezo-actuators. A digital delayed generator was used to synchronously operate all associated instrument, which includes a *d*-DAC, a CCD detector, a high-speed camera, and an oscilloscope. In this study, we used a 1000-seconds long pulse with a 100 ms rise and fall time to the maximum voltage of 150 V (Fig. 1). This particular pulse shape was chosen to match the dynamics of hydrate formation. During the 100 ms rise and fall periods, multiple images and pressures of the sample were obtained *in-situ* using the high-speed camera (solid hexagons in Fig. 1 inset) and time-resolved Ruby luminescence (solid circles in Fig. 1), respectively. The Raman spectra were obtained before and after the pulse (marked as star symbols in the inset).

RESULTS AND DISCUSSION

Figure 2 illustrate the peak pressures of four different experiments (solid symbols) performed in the present study, projected on the phase diagrams of water (solid lines), methane (dash-dotted line), and methane hydrate (dotted lines) [1-4]. At ambient temperatures, methane hydrate crystallizes into cubic structure (MH-I) at <0.1 GPa, which transforms into hexagonal phase II (MH-II) at 0.7 GPa and further to orthorhombic phase III (MH-III) at 1.7 GPa. On the

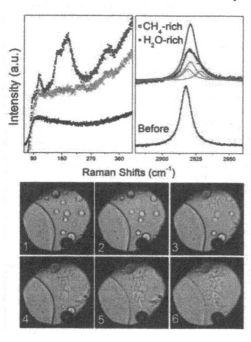

Figure 3. (Upper) Raman spectra of the lattice modes (left) and C-H stretching modes (v_1) of methane and methane hydrates, obtained before and after the peak pressure (stars in Fig. 1). (Bottom) Microphotograph images of the sample showing methane (mostly formed into bubbles) and water (mostly in the transparent area) and methane hydrates (in the area of dendrites between in the image 3 through 6), represented according to the time events as shown in Fig. 1 (hexagons).

89

other hand, water crystallizes into tetragonal ice VI at 0.9 GPa and then cubic ice VII at 2.1 GPa. MH-III and ice VII are stable over a large pressure range of 20-100 GPa well beyond the maximum pressure shown in Fig. 2.

The peak pressures of the present experiments are at: (i) a low pressure of 0.6 GPa in the stability field of water and MH-I, (ii) intermediate pressures of 1.5 GPa and 1.6 GPa in the stability field of ice-VI and MH-II, and (iii) a high pressure of 2.2 GPa in the stability field of ice-VII and MH-III. In the low-pressure experiment, we observe that MH-I forms after the pressure release. This indicates that MH-I forms with an induction time longer than 1000 seconds. In addition, the activation energy associated with the formation of MH-I or the enthalpy of mixing is higher than the energy release of the pressure perturbation.

In the high-pressure experiment, water and methane rapidly solidify into ice-VII and CH_4-I upon the compression, instead of forming methane hydrates. On the other hand, MH-I forms upon the pressure release below 1.0 GPa – the melting point of ice-VI at ambient temperature. This result indicates that methane hydrate forms only from water – not from ice, probably due to slow diffusion rate of methane gas through ice.

In the intermediate-pressure experiments, we observed the formation of MH-II during the compression to 1.5 and 1.6 GPa. Note that in this pressure range the rapidly compressed water at the rate of 0.8-1.6 GPa/s is super-compressed, without solidification, to well within the stability field of ice-VI [11]. Therefore, it is likely that MH-II is formed from super-compressed water (not from ice-VI).

The fact that the hydrate forms from super-compressed water underscores the diffusion controlled process. In this regard, it is interesting to note that methane hydrate forms at the intermediate pressures - not at the low pressure. This is likely due to the fact that methane becomes more miscible with water as pressure increases[12,13], resulting in a faster dynamic of the formation of MH-II than MH-I.

Figure 3 shows the Raman spectra of the sample before and after the pressure pulse and a series of microphotograph images. The Raman signal in the bubble-like area before the pressure increases shows a strong $v_1(CH_4)$ mode at 2917.5 cm^{-1} without any lattice modes – typical spectral characteristics of pure methane [14,15]. The phase separation between water and methane is apparent in the images, consistent with the Raman data. Note that the pressure of sample, ~1.5 GPa, is well beyond the solidification pressure for water (~0.9 GPa) or methane (~0.8 GPa). Yet, there is no apparent crystal facet in the image 1. It indicates that water remains as a super-compressed liquid, while methane transforms into a plastic solid as observed previously [14,15].

Notice in the image 2 that small crystallite-like features grow on the surface of methane-rich bubbles, indicating the formation of methane hydrates. It first appears ~66 sec after the

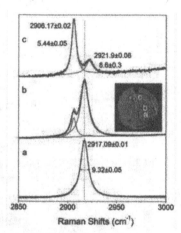

Figure 4. The representative Raman spectra of methane and methane hydrate at 0.7 GPa. The dotted line indicates the peak position of pure methane. The inset and the labels illustrate the corresponding spots of each Raman spectrum.

pressure increases to the peak value; then, grows quickly for the next 30 seconds and joins with nearby methane-rich bubbles (the images **3** to **6**). This interaction indicates the diffusion-controlled growth of methane hydrate. Such interfacial growth of methane hydrate is also evident in the Raman data. For example, the Raman spectrum of CH_4-rich area is characteristic to that of pure methane. The Raman spectrum of H_2O-rich area shows a broad and asymmetry feature centered at ~2920 cm[-1], which can be deconvoluted into three peaks: one for the v_1 mode of methane and two for MH-II. Although the intensity ratio of these two MH-II assigned peaks is not ideal for MH-II, the center of peak positions and the peak splitting of ~10 cm[-1] agree well with those of MH-II. More importantly, the lattice modes are characteristic to those of MH-II[5,6,8].

The broad nature of the Raman spectra in Fig. 3 can be understood in terms of a heterogeneous nature of the sample that is a mixture of methane, water, and methane hydrates. Figure 4 plots the spatially resolved Raman spectra of the sample, which varies strongly depending on the position. For example, the Raman spectra obtained at the position **a** of the image consists of the v_1 of pure methane, indicating it is methane-rich area. The Raman at the position **c** shows a doublet at 2906.2 and 2921.9 cm[-1]. These peak positions and intensity ratio are in perfect agreement with those of MH-I [2,4-6]. The Raman peak at the position **b** appears to be a triplet which can be deconvoluted into the combination of the peak **a** and the doublet **b** as expected at the crystal boundary.

CONCLUSIONS

The most interesting result of the present study is the fact that methane hydrate forms from super-compressed water in the stability field of ice VI and MH-II. The formation of MH-I, on the other hand, seems to accompany an induction period substantially longer than that of MH-II. The fast formation of MH-II is likely due to the enhanced miscibility between super-compressed water and methane at elevated pressures. On the time scale we studied (10 GPa/s), we have not observed the formation of MH-III in its stability field. This is due to the fact that the solidification of super-compressed water occurs much faster than the formation of hydrate above 2 GPa. These experimental observations are consistent with a diffusion controlled growth mechanism for the formation of methane hydrate under pressure.

ACKNOWLEDGMENTS

We appreciate Dr. Minseob Kim in the Institute for Shock Physics at WSU for his assistance of operating the High-pressure loader and thank Dr. Haoyan Wei in the Applied Sciences Laboratory at WSU for helping the modulations of the power supplies. The present study has been supported by the ACS-PRF (49207-ND10) and NSF-DMR (0854618).

REFERENCES

1. Y. A. Dyadic, E. Y. Aledo and E. G. Larine, Mendeleev Communications **7** (1997) 34.

2. T. Kawasaki, Y. Kato, S. Sasaki, T. Kummel and H. Shimizu, Chemical Physics Letters **388** (2004) 18.
3. J. S. Love day, R. J. Names, M. Guthrie, S. A. Belmonte, D. R. Allan, D. D. Klug, J. S. Test and Y. P. Handan, Nature **410** (2001) 661.
4. H. Shimizu, T. Kawasaki, T. Kummel and S. Sasaki, The Journal of Physical Chemistry B **106** (2001) 30.
5. C. Mathieu, M. Yen and G. Olivier, Journal of Raman Spectroscopy **38** (2007) 440.
6. S. Sasaki, Y. Kato, T. Kummel and H. Shimizu, Chemical Physics Letters **444** (2007) 91.
7. M. A. Neumann, W. Press, C. Noldeke, B. Asmussen, M. Prager and R. M. Ibberson, The Journal of Chemical Physics **119** (2003) 1586.
8. J. Korus, G. Firmer, J. Monocle and M. R. Pederson, Modeling and Simulation in Materials Science and Engineering **8** (2000) 403.
9. W. J. Evans, C. S. Yoo, G. W. Lee, H. Cynin, M. J. Lip and K. Vises, Rev. Sci. In strum. **78** (2007) 6.
10. G. W. Lee, W. J. Evans and C.-S. Yoo, Proceedings of the National Academy of Sciences **104** (2007) 9178.
11. G. W. Lee, W. J. Evans and C.-S. Yoo, Physical Review B **74** (2006) 134112.
12. C. W. Blount and L. C. Price, Solubility of methane in water under natural conditions: a laboratory study. Final report, April 1, 1978-June 30, 1982, in "Other Information: Portions of document are illegible" (1982) p. Medium: ED; Size: Pages: 159.
13. A. Chapoy, A. H. Mohammadi, D. Richon and B. Tohidi, Fluid Phase Equilibria **220** (2004) 111.
14. R. Bin, L. Olive, H. J. Jowl and P. R. Salvia, The Journal of Chemical Physics **103** (1995) 1353.
15. S. Lilting, Z. Zhao, A. L. Runoff, C.-S. Zhao and G. Stephan, Journal of Physics: Condensed Matter **19** (2007) 425206.

PEM Fuel Cells and Electrocatalysis

Mater. Res. Soc. Symp. Proc. Vol. 1262 © 2010 Materials Research Society   1262-W05-01

Neutron Imaging Methods for the Investigation of Energy Related Materials: Fuel Cells, Batteries, Hydrogen Storage, and Nuclear Fuel

Eberhard H. Lehmann, Pierre Oberholzer, and Pierre Boillat
Paul Scherrer Institut, CH-5232Villigen PSI, Switzerland

ABSTRACT

Neutron imaging offers several advantages over other imaging methods for the non-invasive studies of materials and operational systems. In this article, we describe recent neutron radiography and tomography results for operational polymer electrolyte membrane fuel cells, gas evolution inside lithium batteries, an evaluation of the sensitivity limits for metal hydride studies, and examples of nuclear fuel inspection. These examples demonstrate that neutron imaging has an important role for investigation of energy-related materials, largely based upon the high neutron scattering cross section of the hydrogen nucleus combined with the near transparency of steel and aluminum. In addition, advances in neutron imaging based upon control of the neutron wavelength, phase coherence, and neutron magnetic spin moment will enable new insights into materials and device studies.

INTRODUCTION

New technologies will play an important role in the world-wide energy supply and delivery Fuel cells and batteries are attractive options for the important issues of mobile and fixed electricity production and storage. The facile storage of hydrogen as an energy carrier is likely to get a higher level of consideration. And, the nuclear energy is widely integrated into our modern life style. Although some of these technologies are already in use, improvements in performance and reliability are needed to enable economic competition with other currently used energy technologies.

This paper is focused on the non-invasive investigation of components and materials for fuel cells, batteries, potential hydrogen storage devices, and nuclear fuel elements. The method is neutron imaging which has been developed and practiced at a high performance level at Paul Scherrer Institute, Switzerland, at the spallation neutron source SINQ. Neutron imaging provides structural information of samples and systems in two dimensions (radiography) and three dimensions (tomography) and is able to study time-dependent phenomena in a quasi-real time regime. One advantage of neutron imaging methods is a high image contrast for the observation of light elements like hydrogen and lithium while at the same time, neutrons can easily penetrate structural materials like aluminum, copper, and even steel. One goal is to investigate polymer electrolyte membrane fuel cells (PEM-FC) to observe water distribution in the membrane region. Different membrane materials are under investigation and the performance is tested under realistic operational conditions while the water distribution is observed. In the case of battery research, two major questions are considered: How is internal gas production related to charge/discharge cycling? Is ion migration visible during battery operation?

Studies in this are have just started and various material combinations will be under evaluation.

Hydrogen storage will become more important when fuel cells and other hydrogen-related energy processes are broadly introduced into society. Gas storage options include high pressure, cryogenics, or chemisorption/physisorption into metals and other chemicals which deliver hydrogen on demand. Materials like zirconium have a high affinity to hydrogen and might be used as storage devices. With the help of neutron imaging methods it is relatively easy to determine the hydrogen uptake and loss under realistic conditions in assemblies of different size.

A final but no less important topic is the investigation of nuclear fuel elements with neutrons in a non-invasive way. It is the advantage of neutron imaging to have a high penetration for uranium while hydrogen accumulation in the fuel cladding can be observed. In this way, the fuel integrity can be checked and the material damage of the zirconium based cladding investigated. The paper will describe the principles of neutron imagine, show the layout of the user facilities, and give examples of the latest investigations.

THE METHOD OF NEUTRON IMAGING

Neutron imaging is a flexible and promising method for non-destructive testing with capabilities both similar and complementary to the more established X-ray imaging methods. Neutron imaging requires a high flux source of thermal or cold neutrons to reach the image quality common in X-ray imaging. Fig. 1 compares the transmission image of an object, a hard disk drive, when imaged with thermal neutrons and X-rays.

Fig. 1: A comparison is shown of neutron (left) and X-ray (right) imaging of a hard disk drive. Neutrons easily penetrate metals, but the neutron flux is attenuated by thin plastic layers due to the high neutron scattering cross section of hydrogen. Conversely, X-rays are attenuated by high-Z elements and can visualize the solder connections points. For samples such as this disk drive, the imaging methods complement each other.

Neutrons can penetrate metals, including high-atomic number Pb and U, better than X-rays while the light elements (H, Li, B, C,...) will attenuate neutrons more than X-rays. In the example in Fig. 1 the solder points are clearly visible in the X-ray image while the thin printed circuit board is only detected with neutrons.

Neutron imaging is nowadays not limited to the radiography mode, but can be applied to three-dimensional inspection with use of various tomography methodologies. Experience shows [1] that the present limit in spatial resolution is on the order of 10 µm and is due to physical constraints from the neutron detection process. More options for neutron imaging can be found in time-dependent studies [2], the use of limited kinetic energy (wavelength) ranges [3], the utilization of the phase properties of spatially coherent neutron beams [4, 5], and studies with magnetic spin polarized neutrons [6].

FACILITIES AT PSI

As the national neutron source for research and application purposes, Paul Scherrer Institut (PSI) has successfully operated the SINQ spallation neutron source since 1997. Until recently [7], SINQ was world's strongest neutron source in power, based upon the spallation process with high energy protons. SINQ compares in neutron flux to a 15 MW research reactor and is well equipped with advanced instruments for neutron scattering and neutron imaging. Fig. 2 shows the SINQ floor plan with the source area on the left and the neutron guide hall on the right side.

There are two permanent installations for neutron imaging available at SINQ: the thermal neutron facility NEUTRA and the cold neutron facility ICON. A third option has been opened recently [8] at the shared beamline BOA, which enables access to neutrons with much longer wavelengths. Table I compares the major parameters at the three beamlines used for neutron imaging.

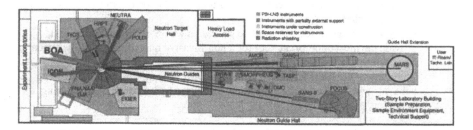

Fig. 2: Layout of SINQ, the Swiss neutron spallation source. Besides the typical neutron scattering devices, the three beamlines for neutron imaging are NEUTRA for thermal neutrons, ICON for cold neutrons and BOA for very cold neutrons

The SINQ neutron imaging beamlines are equipped with state-of-the-art neutron detectors based on locally developed scintillators and cooled-CCD camera sensors. Typical exposure times are on the order of few seconds, depending on the neutron optics selected, which enable time series measurements, tomography sequences, and enhanced methods like phase contrast imaging or energy selective studies. The SINQ neutron imaging facilities are open for external scientific users through a proposal request mechanism coordinated by the PSI user office [9] or on commercial basis.

Table I: Parameters for the Three SINQ Neutron Imaging Beamlines

ICON	NEUTRA	BOA
cold neutrons	thermal neutrons	very cold neutrons
higher contrast	higher penetration	high beam intensity
variable aperture, Bi-filter option	more homogenous illumination for large objects	polarized neutrons
two beam positions	two beam positions	
micro-tomography-setup	two detector boxes	
tilted detector option	X-TRA option (320 kV tube, high current) for referencing	
two detector boxes	option for the inspection of highly activated materials	UNDER CONSTRUCTION
turbine energy selector		
fuel cell infra-structure		

APPLICATIONS IN THE ENERGY FIELD

From the various and numerous applications of neutron imaging, this article concentrates on recent aspects in energy research and the related materials science. As mentioned above, neutrons have the imaging advantage of penetration through thick layers of structural materials like metals while retaining a high sensitivity for organic and other hydrogenous materials. This enables quasi-*in-situ* measurements of processes or the study of the performance properties of energy sources under operation. This article describes the following four examples and gives some of the latest findings:

- Polymer-electrolyte-membrane (PEM) fuel cell, where the combination of hydrogen with oxygen results in liquid or gaseous water, where the liquid phase affects fuel cell performance.
- Lithium-ion batteries, now under intense development, where the gas production during operation can be investigated and probably the ion migration during loading/discharging processes.
- Hydrogen storage in several metallic structures where neutron imaging will provide the most appropriate tool for a non-invasive quantification of hydrogen accumulation in the matrix.
- The study of nuclear fuel to check fuel integrity, the burn-up, and the cladding behavior with respect to the embrittlement caused by hydrogen uptake during operation.

Each of these topics has specific experimental setup requirements and infrastructure needs. The experiments vary with required time and spatial resolution as well as the neutron contrast mechanism. These SINQ experiments are described in more detail below.

Fuel Cell Research

The working principle of a PEM (polymer electrolyte pembrane) fuel cell is show in simplified form in Fig. 3. The two gaseous reagents, hydrogen and oxygen, combine within a membrane where the electro-chemical reaction enables the free electron movement which can be used as primary energy source. Water is the reaction product on the one hand but it is also required to enable the easy charge transfer within the cell. This is the reason why water management plays an essential role in respect to the operational performance and the utilization of the cell under long term, extreme, and transient conditions.

One major potential user of this technology is the automotive industry who intends to replace conventional combustion engines by fuel cell systems in the future. Presently, the existing systems are not yet competitive with respect to price and long term performance (more than five years).

The PSI research facilities have been used for fuel cell studies [10, 11, 12] where the water distribution and the current density have been determined under several external operational conditions such as gas flow, humidity, temperature distribution, and voltage. In-house studies and collaboration with external customers have been driving forces to continuously increase the performance of the experimental setup.

The first studies were done with a relaxed spatial, about 0.2 mm, but relatively high time resolution using test cells to investigate the global water distribution [10, 11]. The latest studies are focused on "differential cells" designed to study the cell behavior near the membrane at highest possible spatial resolution. The 0.2 mm resolution setup is called "trough plane" inspection and is shown in Fig. 4. This setup yields a global overview of where and when liquid water is formed and distributed inside the different fuel cell layers.

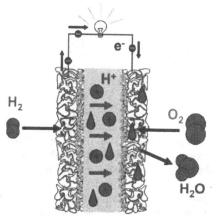

Fig. 3: The operational principle of a PEM fuel cell, based on the electro-chemical combination of hydrogen and oxygen to generate an external current.

For measurements in the membrane region we have to fulfill some additional experimental requirements: The cell must be sufficiently transparent for the neutron beam. The spatial resolution near the membrane is improved by reducing the overall field-of-view. The time resolution is increased by optimizing the neutron scintillator performance.

These objectives drove the creation of operational but simplified "differential cells" with the schematic layout as shown in Fig. 5, together with a photo of the real device in the inspection position. The structural material is mainly Al covered with a thin Au layer for protection against corrosion.

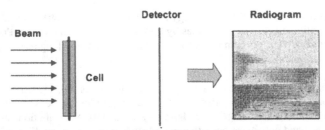

Fig. 4: Through-plane observation of a fuel cell which is placed perpendicular to the neutron beam direction; the observation area is in the order of 10 cm to 30 cm square.

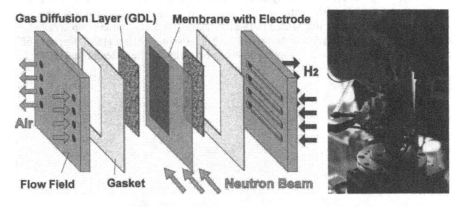

Fig. 5: Differential cell schematic (left), photo (right) of a differential cell in the observation position. The field-of-view during the imaging is on the order of 30 mm delivering the best possible spatial resolution.

The requirement to overcome the recent limitations in spatial resolution was fulfilled in certain respects: a detector setup was designed and built at ICON which is considered to be the best possible, particularly in respect to the optical conditions for the CCD camera. This installation was accompanied with a concerted project to increase the scintillator performance in respect to its inherent resolution, e.g. thickness optimization. A final step has been to use a "tilted option" as depicted in Fig. 6. In this manner, the transmission image was stretched in one direction – across the membrane layer – with the increase of the number of pixels by a factor of 4 or more [12]. The successful application of the device is shown in Fig. 7, where the liquid water distribution is easily visible in the membrane, the gas diffusion layer, and the channels for the gas supply.

Recently, two interesting fuel cell issues have been studied by the PSI Electro-Chemistry Lab at the SINQ neutron imaging beam lines:

1. How to distinguish between the liquid water produced inside the fuel cell by the electrochemical process versus that provided by the humidified gas flow?
2. How the fuel cell operates and reacts in conditions far below the freezing point of water – when the electro-chemical processes begin to suffer from the limited water mobility?

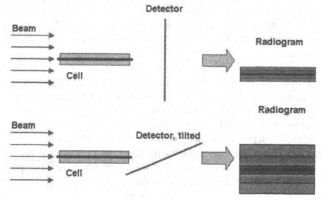

Fig. 6: Two options for the in-plane study of differential fuel cells where the membrane region is precisely aligned parallel to the beam direction. The highest spatial resolution is required for the region of the membrane; this is why a tilted detector option is advantageous compared to the standard perpendicular scintillator orientation.

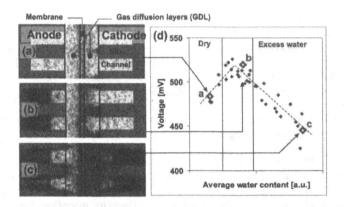

Fig. 7: Results of a PEM fuel cell study as a function of moisture conditions and the measurable voltage. With the tilted detector option, the observation area of 3 mm was stretched by a factor of at least 4 in the horizontal direction. These results clearly demonstrate the high impact of the water phase and distribution on the cell performance.

To answer the first question, we used isotopic substitution in which deuterium (D) instead of hydrogen, in either the hydrogen gas steam or the humidification water. Due to the much smaller attenuation strength for the neutrons of D compared to H, the contrast performance can be used to distinguish between the two water sources. The first such tests have already been performed despite of the relative high price of D_2 and D_2O. The second task has been investigated during operation and imaging of the test cell under controlled thermal conditions around –11 °C. Substantial changes of the operational behavior were found but no clear indication for the ice production can be derived from the image data alone. More sophisticated methods like energy selective imaging techniques are needed to determine the liquid/ice distributions from neutron images.

Li-Ion Batteries

Although Li-ion batteries are already in use for several important technical devises (mobile phones, laptops, cameras) their extensive use is still limited by the reliability, relatively high costs, and low energy density. However, they are considered to be a potential option for a car energy supply if these limitations can be overcome. We think neutron imaging can contribute to battery development and will describe several examples of non-invasive investigations. First, neutron imaging can study electrode materials inside an intact battery during the charge/discharge process. The electrochemical involves Li-ion migration. Due to the fact that ^6Li, the less abundant Li isotope, has a high contrast for neutrons, the sensitivity to detect very small changes in the cell is present. Sensitivity can be enhanced with the use of enriched ^6Li (a strong neutron absorbing media) instead of natural abundance Li.

Another important problem for high performance batteries can be the gas production by and within the electrolyte. Gas imaging studies have been done successfully [13], as shown in Fig. 8. More dedicated studies are needed to determine the optimal beamline and sample parameters for performance studies with either standard batteries or simplified test cells as was done for the PEM fuel cells (see above).

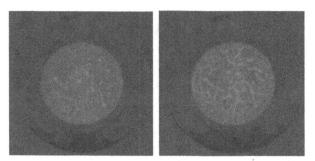

Fig. 8: In-situ study of the gas evolution in a test Li-ion battery with ethylene carbonate and propylene carbonate (2:3, w/w) electrolyte. The two images were taken at 1 h intervals while the voltage was kept stable at 3.3 V [13].

Hydrogen Storage

A reliable and safe hydrogen supply is needed if, e.g., fuel cell technology should be broadly applied. Beside storage as pressurized gas or a cryogenic liquid, hydrogen can also be stored within a metallic matrix. Although this technology was been developed years ago, its introduction in common devices is still pending. Neutron imaging methods are available to make straightforward tests regarding how much, where, and in which manner the hydrogen is stored in the assembly. This is due to the fact that neutrons interact strongly with hydrogen nuclei while the metallic matrix can be mostly transparent. The neutron imaging can also be used for time dependent investigations during storage and discharge of the hydrogen accumulator. We already performed calibration tests with a Zr matrix over the range from 100 to 2000 ppm of hydrogen [14]. The highest sensitivity for hydrogen detection is given by cold neutrons.

Nuclear Fuel Inspection

Nuclear fuel (uranium with enrichment in U-235 on the order of few per cent) is used in the form of pellets filled into cylindrical tubes and assembled into bundled fuel rod elements. The integrity of both the fuel structure and the fuel cladding is an important aspect during the operation of nuclear power plants and afterwards during final disposal.

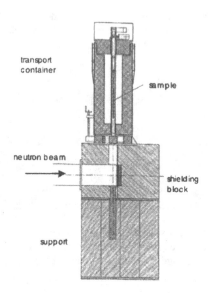

transport
container

sample

neutron beam

shielding
block

support

Fig. 9: The insert of a shielded facility [15] into the NEUTRA beamline enables the non-destructive investigation of highly activated materials like spent nuclear fuel or fuel cladding after long-term radiation exposure.

A non-destructive investigation of the fuel elements is difficult because of the high dose rate of the spent fuel which disables a free access and requires the work to be performed under well shielded conditions (hot laboratory). X-ray methods are impossible to apply due to the very limited transmission of the highly attenuating uranium. Therefore, neutrons are common for such kinds of investigations [15] because of the capability to penetrate uranium easily.

In recent time, some new digital methods were implemented successfully [16] which overcome the previous limitations with film used in the earliest students. Even tomography methods are starting to be available, compatible with the high sample activity. A new setup was built at the NEUTRA facility and implemented for the handling and inspection of irradiated fuel and other heavily activated samples. A diagram is shown in Fig. 9 and the device can be used for fuel rods up to 70 cm in length. The sample can be moved into the beam with 25 cm vertical field of view and rotated precisely about the vertical axis.

Fig. 10: Inspection of a fuel element with pellets of different enrichment where the U-235 content dominates the neutron attenuation contrast. From these image data the fuel integrity and the enrichment can be determined directly.

Fig. 11: The fuel cladding shows hydrogen accumulation (dark areas) at the tube walls after long term radiation exposure, accumulations which might become a source for defects. The neutron inspection enables the non-invasive measurement of the hydrogen content.

Examples for inspection results with this method are shown in Figs. 10 and 11. In the first example, the individual pellets are imaged in high spatial resolution where possible cracks and distortions can be easily observed. Furthermore, the fuel enrichment and the burn-up conditions can be derived quantitatively from the neutron transmission data. The second examples examines the hydrogen content in the fuel cladding which increases during nuclear power plant operation. The probability for cladding defects under normal or accidental conditions increases with hydrogen content. Systematic studies are underway for a better understanding of the phenomena of hydrogen (and oxygen) uptake under the different situations [17].

DISCUSSION

Neutron imaging methods have been demonstrated to be highly important for non-invasive studies of energy relevant materials and systems. The four discussed examples clearly show the high impact of the neutron method for the understanding of the processes and material behavior. In each case a specific setup has to be defined and optimized for the best possible inspection strategy. This holds true with respect to the required precision in the space and time resolution, as well as the effective transmission behavior within the valid detection range of the measurement device. In some cases the safety issues in the handling of activated samples and inflammable reagents must be seriously considered. Compared to past studies with radiography films, the currently used detection methods are able to deliver precise, quantitative information (moisture content, gas production, burn-up, dimensional changes, ...) in a non-invasive manner.

CONCLUSIONS

PSI operates its facilities at the spallation neutron source SINQ for a national and international scientific community on the basis of a proposal system [8]. New explorations are welcomed at the beam lines. More commercially-oriented proposals will be handled preferentially via contracts between the partners.

ACKNOWLEDGMENT

The authors express their thanks to Prof. L. Butler, Louisiana State University, for his strong contribution to make this text readable.

REFERENCES

1. A.S. Tremsin, J.B. McPhate, J.V. Vallerga, O.H.W. Siegmund, J.S. Hull, W.B. Feller, E. Lehmann, Nucl. Instr. and Meth. in Physics Research Section A, Volume 604, Issues 1-2, June 2009, Pages 140-143
2. E.H. Lehmann, Recent improvements in the methodology of neutron imaging, Pramana Journal of Physics, Vol. 71 (No. 4), Oct. 2008, ISSN 0304-4289
3. E.H. Lehmann, G. Frei, P. Vontobel, L. Josic, N. Kardjilov, A. Hilger, W. Kockelmann, A.Steuwer, Nucl. Instr. and Meth. in Physics Research Section A: Volume 603, Issue 3, 21 May 2009, Pages 429-438
4. N. Kardjilov, E. Lehmann, E. Steichele, P. Vontobel, Nucl. Instr. And Meth. A 527 (2004) 519
5. F. Pfeiffer, C. Grünzweig, O. Bunk, G. Frei, E. Lehmann, C. David, Phys. Rev. Lett. 96, 215505 (2006)
6. M. D, I. Manke, N. Kardjilov, A. Hilger, M. Strobl, J. Banhart, New Journal of Physics 11 (2009) 043013
7. http://neutrons.ornl.gov/diagnostics/channel13/ch14.html
8. U. Filges, Redesign of the neutron optical instrument FUNSPIN into Silenos, Accepted proposal to the Swiss National Science Foundation, May 15th, 2009
9. http://user.web.psi.ch/
10. D. Kramer, J. Zhang, R. Shimoi, E. Lehmann, A. Wokaun, K. Shinohara, G. Scherer, Electrochimica Acta 50, 2603 (2005)
11. E. Lehmann, G. Frei, G. Kühne, P. Boillat, Nucl. Instr. And Meth. A 576 (2-3), (2007) 389
12. P. Boillat, G. Frei, E. H. Lehmann, G. G. Scherer, and A. Wokaun, Electrochemical and Solid-State Letters, 13 (3) B25-B27 (2010)
13. D. Goers, M. Holzapfel, W. Scheifele, E.Lehmann, P. Vontobel, P. Noak, J. Power Sources, Vol. 130, pp. 221-226 (2004)
14. E. Lehmann, P. Vontobel, N. Kardjilov, Hydrogen distribution measurements by neutrons, APPLIED RADIATION AND ISOTOPES 61 (4): 503-509 OCT 2004
15. E. Lehmann, P. Vontobel., L. Wiezel, The investigation of highly activated samples by neutron radiography at the spallation source SINQ, Nondestr. Test. Eval. Vol. 16, pp. 203-214, **(YEAR-YEAR???)**
16. M. Tamaki et al., Nucl. Instr. & Meth. in Phys. Res. A 542 (2005) 320-323
17. M.Grosse, E. Lehmann, P. Vontobel, M. Steinbrueck, Nucl. Instr. and Meth. in Phys. Res. 566 (2), 739–745, (2006).

Poster Session

Mater. Res. Soc. Symp. Proc. Vol. 1262 © 2010 Materials Research Society   1262-W06-03

Electronic Structure of La(Fe$_{0.88}$Si$_{0.12}$)$_{13}$

Nozomu Kamakura [1], Tetsuo Okane [1], Yukiharu Takeda [1], Shin-ichi Fujimori [1], Yuji Saitoh [1], Hiroshi Yamagami [1,2], Atsushi Fujimori [1,3], Asaya Fujita [4], Shun Fujieda [5], Kazuaki Fukamichi [5]

[1] Synchrotron Radiation Research Center, Japan Atomic Energy Agency, Hyogo 679-5148, Japan
[2] Department of Physics, Kyoto Sangyo University, Kyoto 603-8555, Japan
[3] Department of Physics, University of Tokyo, Tokyo 113-0033, Japan
[4] Department of Materials Science, Graduate School of Engineering, Tohoku University, Sendai 980-8579, Japan
[5] Institute of Multidisciplinary Research for Advanced Materials, Tohoku University, Sendai 980-8577, Japan

ABSTRACT

La(Fe$_{0.88}$Si$_{0.12}$)$_{13}$ shows peculiar magnetic properties such as the first order paramagnetic-ferromagnetic transition and magnetic-field induced metamagnetic transition accompanied by the lattice expansion. The practical application using the magnetic transition temperature controlled by hydrogen absorption is expected in this compound. Here, the electronic structure of La(Fe$_{0.88}$Si$_{0.12}$)$_{13}$ has been investigated by photoemission spectroscopy using synchrotron soft x-rays. The Fe 3s core-level photoemission spectra below and above the Curie temperature T_C exhibit a satellite structure at ~ 4.3 eV higher binding energy than the main peak, which is attributed to the exchange splitting due to the local moment of Fe. The exchange splitting of the Fe 3s photoemission spectrum with the asymmetric line shape shows that the magnetization of La(Fe$_{0.88}$Si$_{0.12}$)$_{13}$ is derived by the exchange split Fe 3d bands like the itinerant ferromagnetism in Fe metal, while the magnetic transition of La(Fe$_{0.88}$Si$_{0.12}$)$_{13}$ is the first order. The valence band photoemission spectrum shows temperature dependence across the T_C. The temperature dependence of the photoemission spectra is discussed based on the difference between the electronic structure in the ferromagnetic phase and that in the paramagnetic phase.

INTRODUCTION

La(Fe$_x$Si$_{1-x}$)$_{13}$ shows a variety of magnetic properties depending on the composition [1-9]. The ferromagnetic state is stabilized in a wide range of Fe content. With increasing Fe content, the Curie temperature T_C decreases and the ferromagnetic transition changes from a second order to a first order at x = 0.86. In the composition x > 0.86, the magnetic field-induced metamagnetic transition takes place between the critical temperature T_0 and the T_C. Characteristic of La(Fe$_x$Si$_{1-x}$)$_{13}$ is to absorb hydrogen. The lattice expansion by hydrogen absorption in La(Fe$_x$Si$_{1-x}$)$_{13}$ increases the T_C up to room temperature [7-9]. The magnetic transition involving the lattice expansion enables one to apply La(Fe$_x$Si$_{1-x}$)$_{13}$H$_y$ to large magnetostrictive materials and magnetic refrigerator using magnetocaloric effect that can be controlled around room temperature.

In this study, the electronic structure of La(Fe$_x$Si$_{1-x}$)$_{13}$ across the first order magnetic transition is investigated by photoemission spectroscopy. Since the first order magnetic transition maintains even after the hydrogen absorption, the valence bands in the paramagnetic and ferromagnetic phases are studied in La(Fe$_x$Si$_{1-x}$)$_{13}$ with the composition x = 0.88 before hydrogen absorption.

The structure of $La(Fe_{0.88}Si_{0.12})_{13}$ is the cubic $NaZn_{13}$-type structure as shown in figure 1. The Fe and Si atoms are located at the 8b sites or 96i sites forming the icosahedron [4]. A large amount of Fe atoms forms icosahedral clusters which have a local symmetry similar to that in a fcc structure. The Fe-derived states hybridize with Si states, which is considered to control the magnetic moment of $La(Fe_xSi_{1-x})_{13}$ depending on the composition x. In this study, the local magnetic moment of Fe, which is considered to derive the magnetic behavior of $La(Fe_{0.88}Si_{0.12})_{13}$, is investigated by the Fe 3s photoemission spectroscopy.

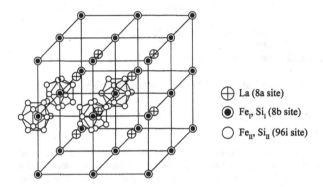

⊕ La (8a site)

◉ Fe_I, Si_I (8b site)

○ Fe_{II}, Si_{II} (96i site)

Figure 1. Crystal structure of $La(Fe_{0.88}Si_{0.12})_{13}$.

EXPERIMENT

Photoemission experiment was performed at BL23SU of SPring-8 equipped with a SCIENTA SES2002 analyzer. The sample surface was obtained by fracture in the ultra high vacuum chamber. The T_C of $La(Fe_{0.88}Si_{0.12})_{13}$ is reported to be 195 K [1-4]. The electronic states in the ferromagnetic and paramagnetic phases are measured at 180 K and 220 K, respectively. The valence band photoemission spectra have been measured with $hv = 800$ eV and the core-level photoemission spectra have been measured with $hv = 1253.6$ eV.

RESULTS and DISCUSSION

Figure 2(a) shows the valence band photoemission spectra for $La(Fe_{0.88}Si_{0.12})_{13}$ in the ferromagnetic phase measured at 180 K. The spectrum in the ferromagnetic phase shows the main peak at 1.2 eV and additional broad structure at ~ 2.5 eV. The reported calculation has predicted that the majority spin band of $La(Fe_{0.88}Si_{0.12})_{13}$ is fully occupied in the ferromagnetic phase [3]. Thus, the peak near the Fermi level (E_F) in the density of states (DOS) for the majority spin band is located at the relatively high binding energy ~ 1.2 eV, which coincides with the peak position of the minority spin DOS [3]. There is another peak at ~ 2.6 eV in the majority spin DOS of the reported calculation [3]. The energy positions of the main peak and broad structure in figure 2(a) are consistent with those in the calculated DOS for the ferromagnetic phase [3], indicating the fully occupied Fe $3d$ majority spin bands with the magnetization of 2.0 μ_B.

The valence band photoemission spectrum shows the temperature dependence across the T_C (figure 2(b)). This temperature dependence can be considered to originate from the difference of the electronic states between the ferromagnetic and paramagnetic phases. The reported band calculation has indicated that there is a DOS peak at the E_F in the paramagnetic phase, while the valence band in the ferromagnetic phase shows less intensity near the E_F due to the fully occupied majority spin band [3]. In the paramagnetic calculation, there are also DOS structures around 1.7 eV, in which the valley of the two DOS peaks is situated in the ferromagnetic phase [3]. In figure 2(b), the measured photoemission spectrum in the paramagnetic phase shows an additional intensity at the E_F and a broad feature on the high binding energy side of the main peak, in comparison with the spectrum of the ferromagnetic phase. These changes in the valence band photoemission spectrum agree with the difference between the calculated DOSs in the paramagnetic and ferromagnetic phases [3].

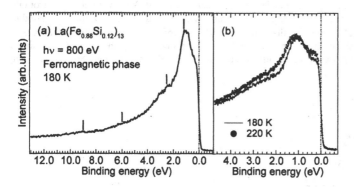

Figure 2. (a) Valence band photoemission spectrum measured at 180 K (below the T_C). Peaks in the spectrum are marked with bars. (b) Photoemission spectra near the E_F at 180 K (below the T_C) and 220 K (above the T_C) shown by solid line and dots, respectively.

So far, the temperature dependence of the valence band photoemission spectrum during the paramagnetic-ferromagnetic transition has been extensively investigated in the $3d$ transition metal ferromagnets Fe and Ni, in which the validity of the Stoner-like behavior across the T_C was examined [10-13]. The recent angle resolved photoemission spectroscopy (ARPES) studies have indicated that the exchange splitting of the Ni $3d$ band gradually decreases with the increase of temperature due to a second order phase transition from the ferromagnetic phase to the paramagnetic phase, showing Stoner-like behavior [11-13]. In contrast to the usual itinerant ferromagnet, the magnetization of La(Fe$_{0.88}$Si$_{0.12}$)$_{13}$ exhibits a discontinuous change at the T_C due to the first order phase transition [1-4]. Accordingly the temperature dependence of the valence band photoemission spectrum shows a difference in the narrow temperature range between 180 K and 220 K. Thus, the temperature dependence originating from the exchange split valence band and the collapse of the exchange splitting is clarified in the DOS obtained by the valence band photoemission of La(Fe$_{0.88}$Si$_{0.12}$)$_{13}$ across the first order magnetic transition.

Figure 3(a) shows the Fe 3*s* core-level photoemission spectrum for La(Fe$_{0.88}$Si$_{0.12}$)$_{13}$. The Fe 3*s* spectrum shows a satellite structure at ~ 4.3 eV higher binding energy than the main peak. In several magnetic 3*d* transition metal compounds, the satellite structure of the 3*s* photoemission spectra on the high binding energy side has been attributed to the exchange splitting that reflects the local magnetic moment [14-16]. The Mössbauer result indicates that the magnetization of La(Fe$_x$Si$_{1-x}$)$_{13}$ is derived from the Fe magnetic moment [5, 6], while the exchange splitting in the 3*s* spectrum is the local probe of the magnetic moment insensitive to the long range magnetic order [5, 6]. In figure 3(a), thus, the satellite structure can be considered to originate from the exchange splitting of the Fe 3*s* spectrum. Under the assumption of constant exchange integrals for a given element, the exchange splitting of the 3*s* spectrum is expected to reveal the linear dependence on the local magnetic moment [14, 15, 17]. The previous reports on the 3*s* core-level photoemission of several Fe and Co compounds show almost linear dependence between the exchange splitting and magnetic moment [14, 15], though the estimation from the 3*s* core-level spectrum sometimes involves latent ambiguity [16]. When the linear relation reported in [15] is applied, the local moment of Fe in La(Fe$_{0.88}$Si$_{0.12}$)$_{13}$ can be estimated to be about 2.3 μ_B from the exchange splitting of the Fe 3*s* spectrum. This value is close to the magnetization estimated by SQUID (2.0 μ_B) [2], showing that the Fe 3*d* electrons derive the magnetization of La(Fe$_{0.88}$Si$_{0.12}$)$_{13}$. As shown in figure 3(a), the Fe 3*s* spectra below and above the T_C are almost identical, indicating the existence of almost constant Fe local moment regardless of the long range magnetic order.

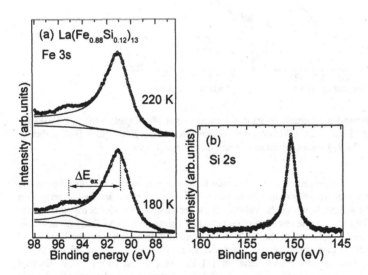

Figure 3. (a) Fe 3*s* core-level photoemission spectra for La(Fe$_{0.88}$Si$_{0.12}$)$_{13}$ at 180 K and 220 K. (b) Si 2*s* core-level photoemission spectrum in the ferromagnetic phase.

In the valence band photoemission spectrum of figure 2(a), the additional spectral features are observed at high binding energies (~ 6.0 and 9.0 eV). In the Fe-Si compounds, the hybridization with the Si derived state causes the reduction of the Fe magnetic moment and the Si-derived states are located on the high binding energy side of the Fe $3d$ state [18, 19]. The Si $3p$ and $3s$ states have been reported to be located at 4.4 and 9.4 eV in Fe_3Si, respectively [18]. In $La(Fe_xSi_{1-x})_{13}$, the role of the Si-derived states in the ferromagnetic behavior has been investigated and the addition of Si has been considered to also cause the reduction of the Fe magnetic moment by the hybridization. As discussed in [6], the substitution of Si in $La(Fe_xSi_{1-x})_{13}$ brings about the reduction of the Fe $4s$ and $3d$ occupied electrons. The Fe $3d$ minority spin bands overlap with the Si sp down spin bands, which lowers their energies with respect to the Fe $3d$ majority spin bands and causes the reduction of the Fe magnetic moment in the high Si concentration [6]. In the valence band spectrum for $La(Fe_{0.88}Si_{0.12})_{13}$ of figure 2(a), the peaks at high binding energies (~ 6.0 and 9.0 eV) can be attributed to the Si $3p$ and Si $3s$ states. Through the hybridization, the Si atoms reduce the net charge densities of Fe $4s$ electrons [6]. Accordingly, the line shape of the Si $2s$ core-level spectrum is symmetric with the asymmetric parameter of 0.03 (figure 3(b)) and that of the Fe core-level spectra is asymmetric (the asymmetric parameter of 0.2 for the Fe $3s$ spectra (figure 3(a))) [20]. These line shapes of the core-level spectra suggest that the effects of metallic screening on Si core-levels are small, which is similar to the Si $1s$ core-level spectra for $Fe_{3-x}V_xSi$ [21].

CONCLUSIONS

In this study, the electronic structure of $La(Fe_{0.88}Si_{0.12})_{13}$ is investigated. The valence band photoemission spectrum near the E_F below the T_C is consistent with the reported calculation for the ferromagnetic phase. The temperature dependence of the valence band photoemission spectrum shows temperature dependence across the T_C, which is attributed to the difference of the electronic states between the ferromagnetic and paramagnetic phases. The Fe $3s$ core-level photoemission spectrum shows the exchange splitting due to the existence of the Fe local moment regardless of the presence of long-rang magnetic order. As found from the photoemission spectroscopy, the Fe local moment in $La(Fe_{0.88}Si_{0.12})_{13}$ is almost constant below and above the T_C and the exchange splitting of the Fe $3d$ bands occurs due to the ferromagnetic spin order below the T_C.

REFERENCES

1. A. Fujita, Y. Akamatsu, and K. Fukamichi, J. Appl. Phys. **85**, 4756 (1999).
2. A. Fujita, S. Fujieda, K. Fukamichi, H. Mitamura, and T. Goto, Phys. Rev. B **65**, 014410 (2001).
3. A. Fujita, K. Fukamichi, J.-T. Wang, and Y. Kawazoe, Phys. Rev. B **68**, 104431 (2003).
4. S. Fujieda, A. Fujita, K. Fukamichi, Y. Yamaguchi, and K. Ohoyama, J. Phys. Soc. Jpn. **77**, 074722 (2008).
5. X. B. Liu, Z. Altounian, and D. H. Ryan, J. Phys.: Condens. Matter **15**, 7385 (2003).
6. H. H. Hamdeh, H. Al-Ghanem, W. M. Hikal, S. M. Taher, J. C. Ho, D. T. K. Anh, N. P. Thuy, N. H. Duc, and P. D. Thang, J. Magn. Magn. Mater. **269**, 404 (2004).
7. S. Fujieda, A. Fujita, K. Fukamichi, Y. Yamazaki, and Y. Iijima, Appl. Phys. Lett. **79**, 653 (2001).
8. A. Fujita, S. Fujieda, Y. Hasegawa, and K. Fukamichi, Phys. Rev. B **67**, 104416 (2003).

9. J. Lyubina, K. Nenkov, L. Schultz, and O. Gutfleisch, Phys. Rev. Lett. **101**, 177203 (2008).

10. E. Kisker, K. Schröder, M. Campagna, and W. Gudat, Phys. Rev. Lett. **52**, 2285 (1984).

11. P. Aebi, T. J. Kreutz, J. Osterwalder, R. Fasel, P. Schwaller, and L. Schlapbach, Phys. Rev. Lett. **76**, 1150 (1996).

12. T. Greber, T. J. Kreutz, and J. Osterwalder, Phys. Rev. Lett. **79**, 4465 (1997).

13. T. J. Kreutz, T. Greber, P. Aebi, and J. Osterwalder, Phys. Rev. B **58**, 1300 (1998).

14. D. G. Van Campen and L. E. Klebanoff, Phys. Rev. B **49**, 2040 (1994).

15. F. Bondino, E. Magnano, M. Malvestuto, F. Parmigiani, M. A. McGuire, A. S. Sefat, B. C. Sales, R. Jin, D. Mandrus, E. W. Plummer, D. J. Singh, and N. Mannella, Phys. Rev. Lett. **101**, 267001 (2008).

16. J. F. van Acker, Z. M. Stadnik, J. C. Fuggle, H. J. W. M. Hoekstra, K. H. J. Buschow, and G. Stroink, Phys. Rev. B **37**, 6827 (1988).

17. J. H. Van Vleck. Phys. Rev. **45**, 405 (1934).

18. B. Egert and G. Panzner, Phys. Rev. B **29**, 2091 (1984).

19. F. Sirotti, M. D. Santis, and G. Rossi, Phys. Rev. B **48**, 8299 (1993).

20. S. Doniach and M. Šunjić, J. Phys. C **3**, 285 (1970).

21. Y. T. Cui, A. Kimura, K. Miyamoto, M. Taniguchi, T. Xie, S. Qiao, K. Shimada, H. Namatame, E. Ikenaga, K. Kobayashi, H. Lin, S. Kaprzyk, A. Bansil, O. Nashima, and T. Kanomata, Phys. Rev. B **78**, 205113 (2008).

Mater. Res. Soc. Symp. Proc. Vol. 1262 © 2010 Materials Research Society   1262-W06-09

Ultrahigh-pressure study on the magnetic state of iron hydride using an energy domain synchrotron radiation [57]Fe Mössbauer spectrometer

Takaya Mitsui [1,2] and Naohisa Hirao [3]

[1] Japan Atomic Energy Agency, 1-1-1 Kouto, Sayo-cho, Sayo-gun, Hyogo 679-5148, Japan

[2] CREST, Japan Science and Technology Agency, 4-1-8 Honcho, Kawaguchi, Saitama 332-0012, Japan

[3] Japan Synchrotron Radiation Research Institute, 1-1-1 Kouto, Sayo-cho, Sayo-gun, Hyogo 679-5198, Japan

ABSTRACT

Pressure induced magnetic phase transition of iron hydride was investigated with an in-situ Mössbauer spectrometer using synchrotron radiation (SR). The spectrometer is composed of a high resolution monochromator, an X-ray focusing device, a variable frequency nuclear monochromator and a diamond anvil cell. The optical system, advantages of the spectrometer and the observed high pressure magnetic phases of iron hydride are described.

INTRODUCTION

Fundamental research of metal hydride under high hydrogen pressure is a very important for understanding hydrogen-induced phenomena in metallic materials. Since the dissolution of hydrogen into metal under high pressure gives rise to some remarkable changes in physical properties. As a typical phenomenon, at the hydrogen pressure of 3.5 GPa in room temperature, the iron transforms from the body centered cubic (bcc) phase of Fe to the double hexagonal close packing (dhcp) phase of FeH_x with two different hexagonal and cubic iron sites [1]. The formed dhcp structure is stable up to at least 80 GPa in room temperature [2]. Compared to pure iron, iron hydride shows some unique physical properties [1,3-9]. As a remarkable feature, the dhcp-FeH_x is ferromagnetic, in contrast to the nonmagnetic hcp-Fe [10,11]. Since magnetism is, in general, closely related to volume expansion, the hydrogen-induced volume expansion probably contributes to the appearance of ferromagnetic dhcp-FeH_x. Some experimental [2,9] and theoretical [12-14] studies of the ferromagnetic dhcp-FeH_x predict magnetic phase transition at high pressure due to the decrease in atomic volume. However, the magnetic transition process has not been observed clearly. Therefore, we carried out in-situ high-pressure synchrotron radiation Mössbauer spectroscopy (hp-SRMS) of iron hydride using a diamond anvil cell. The magnetic states of dhcp-FeH_x were probed by a newly developed high pressure synchrotron radiation (SR) [57]Fe-Mössbauer spectrometer [15,16]. In the present paper, we repot the optical system, potential benefits of the hp-SRMS and the observed ferromagnetic to nonmagnetic transition of dhcp-FeH_x at high-pressure conditions up to about 65 GPa.

MEASUREMENT SYSTEM OF ENERGY DOMAIN HP-SRMS

In-situ Mössbauer spectroscopy (MS) with a diamond anvil cell (DAC) is a suitable tool for the study of pressure-induced effects on the magnetism of solid materials. However, MS study under high hydrogen pressure need to fill a sufficient liquid hydrogen-reservoir to a small sample chamber in a DAC. This condition limits the available sample size compared with the usual high-pressure MS experiments. As a result, as far as we know, the conventional MS with a radioactive isotope (RI) has not achieved the high hydrogen pressure study beyond 50 GPa; the measurement technique on the micron-sized sample in a DAC is essential for such a study.

To solve the problem, we have recently developed a new SR based Mössbauer spectrometer, which allows us to measure the energy domain ^{57}Fe-high-pressure SR Mössbauer spectroscopy using the DAC of micron-scale gasket-hole size. It has been installed at the BL11XU of SPring-8. The main optical components are a high resolution monochromator (HRM), a multilayer X-ray focusing mirror (MXFM) and a variable frequency nuclear monochromator (VFNM) using a ^{57}FeBO$_3$ single crystal [16]. The external view and optical system of the hp-SRMS are shown in figure 1(a) and 1(b), respectively.

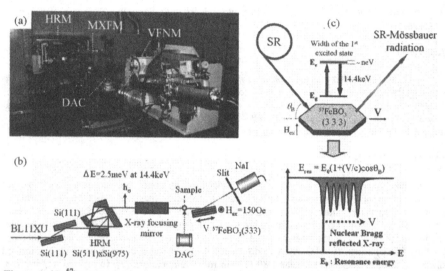

Figure 1. SR ^{57}Fe-Mössbauer spectrometer for ultrahigh-pressure experiment.
(a) External view of a hp-SRMS spectrometer (b) Optical system (c) Conceptual diagram of the energy domain SR ^{57}Fe-Mössbauer spectroscopy with a VFNM.

The measurement procedure of hp-SRMS is as follows. The electron current of the storage (SPring-8) ring was operated at 100 mA at 8.0 GeV. A liquid-nitrogen-cooled Si(111) double-crystal monochromator was used to handle the high heat load of undulator radiation. In order to reduce the huge electronic Tommson scattering noise, a σ-polarized incident X-rays was pre-monochromatized with an energy width of 2.5 meV at 14.4 keV nuclear resonance of ^{57}Fe nuclei by a nested high resolution monochromator consisting of asymmetric Si (5 1 1) and asymmetric Si (9 7 5) channel-cut crystals. The beam size was 0.4 mm x 1.8 mm and the total flux was 1.0 x 10^{10} cps. A small-size probe beam was obtained by a bent elliptical multilayer X-ray focusing mirror, which was coated with 50 layers of [W (13 Å) / Si (39.5 Å)] on a high-quality quartz base. Here, the incident X-rays were reflected at a Bragg peak angle of 8.6 mrad and were horizontally focused with a size of 400 μm x 20 μm at the focus position of 600 mm downstream from the center poison of the focusing mirror. The total flux was 4.5 x 10^9 cps. The focused X-rays were ultrafinely monochromatized to the bandwidth of 15.4 neV by an $^{57}FeBO_3$ (3 3 3) pure nuclear Bragg reflection (PNBR) near the Néel temperature (75.8 °C) in a 150 Oe external field. Here, the PNBR forbade the electronic Tommson scattering, but it allowed the nuclear resonant scattering of ^{57}Fe nuclei (SR-Mössbauer radiation). The narrow bandwidth of neV order was achieve by the magnetically collapse of the PNBR near the Néel temperature in the presence of a small magnetic field [17]. Behind a slit, the nuclear Bragg diffracted X-rays were detected by a NaI (Tl) detector. The typical peak photon counting rates was 3.5×10^3 cps and the noise level was below 4.0 %. Moreover, as shown in figure 1(c), the resonance energy of PNBR was Doppler shifted by oscillating the crystal parallel to the reflection plane in the sinusoidal velocity mode with the frequency of 10 Hz. In this optics, when the sample enclosed in the DAC was placed at the beam focus position, the MS spectrum was measured by counting the reflection intensity as a function of the velocity.

RESULTS AND DISCUSSION

In the present experiment, we carried out the ^{57}Fe hp-SRMS on the iron hydride under high hydrogen pressures up to 64.7 GPa. As shown in figure 2(a), a small polycrystalline iron metal (^{57}Fe 95 %) and ruby pressure markers were enclosed in the DAC filled with liquid hydrogen at pressure of 180 MPa at 300K. Here, hydrogen acts as both a hydrogen reservoir for the iron-hydrogen reaction and a pressure-transmitting medium. Figure 2(b) presents the observed room temperature SR ^{57}Fe-Mössbauer spectra for different pressures. Each spectrum is measured with a considerable short measurement time of 5.0 hours. One should note that all spectra are obtained with good statistical quality in spite of the very small sample (< φ50 μm). It proves a special advantage of hp-SRMS on the high hydrogen pressure study using a DAC.

At hydrogen-inert pressure of 1.8 GPa, the spectrum shows a typical absorption sextet corresponding to the well split hyperfine structure of pure bcc-Fe. In contrast, at P = 7.9 GPa, iron hydride is formed and the spectrum shows the two well-resolved nuclear Zeeman sextets owing to hexagonal and cubic iron sites. With the build up of pressure, the hyperfine field shows a gradual decrease up to 24.7 GPa and a rapid break down at 27.6 GPa. On the other hand, the residual weak hyperfine field decreases sluggishly with the buildup of pressure (P > 27.6 GPa) and finally disappears at 64.7 GPa. This result is clear evidence of the pressure-induced phase transition from the ferromagnetic to nonmagnetic state in dhcp-FeH$_x$. Here, the rapid break down of the hyperfine field of dhcp-FeH$_x$ is qualitatively understood in the frame of Storner theory for itinerant magnetism [18]. In this scheme, a lattice expansion is induced by the iron hydride formed in low-pressure phase (P ≥ 3.5 GPa) and it leads a narrowing of the electronic band width. In such a case, the density of state at the Fermi level is enhanced so that the Stoner condition for ferromagnetism is satisfied because the molecular field coefficient is not sensitive to pressure. As a result, the Mössbauer spectrum shows a hyperfine splitting profile of a ferromagnetic dhcp-FeH$_x$. However, in high-pressure phase (P ≥ 27.6 GPa), the lattice compression leads to a broadening of electronic band width [19]. Then, the density of state at the Fermi level is decreased as compared with the low pressure phase, so that the Stoner condition is no longer satisfied. Consequently, MS spectrum shows a single line absorption profile corresponding to the non-magnetic state in dhcp-FeH$_x$. The broadening to flat Fe 3d band of dhcp-FeH$_x$ has been experimentally observed by Mao et al. in nuclear resonant inelastic X-ray scattering experiments at high pressures [9].

In the present work, the detailed magnetic behaviors, including the sluggish feature in the high pressure phase, are not clear yet. Such a phenomenon may be related to the different pressure effect on the magnetic state of the iron atoms belonging to hexagonal and cubic sites. Currently, we are preparing further high-pressure measurements, including the SR-XRD and hp-SRMS at low temperature, to study the correlation between structural and magnetic properties on the high-pressure phase of dhcp-FeH$_x$.

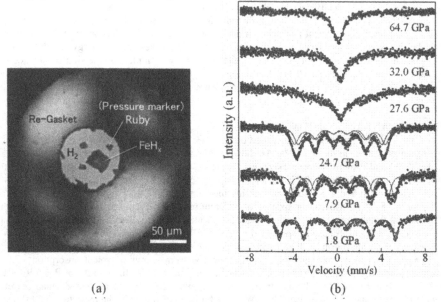

(a)　　　　　　　　　　　　　　(b)

Figure 2. SR Mössbauer spectroscopy of FeH$_x$. (a) Sample conditions at 1.8 GPa (b) Mössbauer spectra of iron under high hydrogen pressures up to 64.7 GPa. Solid and dashed lines correspond to calculations.

SUMMARY

The magnetic state of dhcp-FeH$_x$ under the ultrahigh pressure up to about 65GPa was studied by a new hp-SRMS spectrometer. In the high pressure phase, the measured MS spectra showed clear single-line absorption profiles suggesting the ferromagnetic to nonmagnetic phase transition of dhcp-FeH$_x$. The hyperfine field shows a rapid break down at 24.7 GPa and a sluggish reduction with the further buildup of pressure (P > 24.7 GPa). The magnetic phase transition of dhcp-FeH$_x$ is mainly due to the unique volume effects competed by hydrogenation (expansion) and high-pressure (compression).

ACKNOWLEDGMENTS

The authors would like to thank Dr. Y. Ohishi and Dr. K. Aoki for various fruitful discussions and continuously encouragement. The authors are also grateful to Professor M. Seto and Professor H. Yamagami for various worthwhile comments. This work was supported by the New Energy and Industrial Technology Development Organization (NEDO) under the Advanced Fundamental Research Project on Hydrogen Storage Materials.

REFERENCES

1. J. V. Badding, R. J. Hemley and H. K. Mao, Science **253**, 421 (1991).
2. N. Hirao, T. Kondo, E. Ohtani, K. Takemura and T. Kikegawa, Geophys. Res. Lett. **31**, L06616 (2004).
3. Y. Fukai, K. Mori and H. Shinomiya, J. Alloys Comp. **348**, 105 (2003).
4. V. E. Antonov, I. T. Belash and E. G. Ponyatovsky, Scripta Met. **16**, 203 (1982).
5. V. E. Antonov, I. T. Belash, E . G. Ponyatovskii, V. G. Thiessen and V. I. Shiryaev, Phys. Stat. Sol. (a) **65**, K43 (1981).
6. R. Wordel, F. E. Wagner, V. E. Antonov, E. G. Ponyatovskii, A. Permogorov, A. Planchinda and E. F. Makarov, Hyperfine Interact. **28**, 1005 (1986).
7. I. Choe, R. Ingalls, J. M. Brown, Y. Sato-Sorensen and R. Mills, Phys. Rev. B **44**, 1 (1991).
8. G. Schneidera M. Baier, R. Wordel, F.E. Wagner, V.E. Antonov, E.G. Ponyatovsky, Yu. Kopilovskii and E. Makarov, J. Less-Common Met. **172–174**, 333 (1991).
9. W. L. Mao, W. Sturhahn, D. L. Heinz, H. Mao, J. Shu, and R. J. Hemley, Geophys. Res. Lett. **31**, L15618 (2004).
10. R. D. Taylor, M. P. Pasternak and R. Jeanloz, J. Appl. Phys. **69**, 6126 (1991).
11. S. Nasu, T. Sasaki, T. Kawakami, T. Tsutsui and S. Endo, J. Phys.: Condens. Matter **14** (2002) 11167.
12. C. Elsässer, Jing Zhu, S. G. Louie , B. Meyer, M. Fähnle and C. T. Chan, J. Phys. Condens. Matter **10**, 5113 (1998).
13. M. E. Pronsato, G. Brizuela and A. Juan, J. Phys. Chem. Solids **64**, 593 (2003).
14. A. S. Mikhaylushkin, N. V. Skorodumova, R. Ahuja and B. Johansson, AIP Conf. Proc. **837**, 161 (2006).
15. T. Mitsui, M. Seto, N. Hirao, Y. Ohishi and Y. Kobayashi, Jpn. J. Appl. Phys. **46**, L382 (2007).
16. T. Mitsui, N. Hirao, Y. Ohishi, R. Masuda, Y. Nakamura, H. Enoki, K. Sakaki and M. Seto, J. Synchrotron Radiat. **16**, 723 (2009).
17. G. V. Smirnov, M. V. Zelepukhin and U. Van Bürck, JETP Lett. **43**, 352 (1986).
18. E. C. Stoner, Proc. Royal Soc. London A **165**, 372 (1938).
19. V. Heine, Phys. Rev. **153**, 673 (1967).

Mater. Res. Soc. Symp. Proc. Vol. 1262 © 2010 Materials Research Society   1262-W06-10

Real-time and Direct Observation of Hydrogen Absorption Dynamics for Pd Nanoparticles

Daiju Matsumura, Yuka Okajima, Yasuo Nishihata and Jun'ichiro Mizuki
Synchrotron Radiation Research Center, Japan Atomic Energy Agency, 1-1-1 Koto, Sayo, Hyogo 679-5148, Japan

ABSTRACT

Dynamic local structural change of Pd nanoparticles on alumina surface during hydrogen absorption process was directly observed by x-ray absorption fine structure spectroscopy with dispersive mode. Main four parameters of x-ray absorption spectroscopy were determined even in the case of 50 Hz observation. It is clearly revealed that Pd nanoparticles directly change to the hydride phase in 50 ms at 200 kPa of hydrogen pressure. Although large lattice expansion was observed, significant structural distortion was not investigated in the results of the change of Debye-Waller factor.

INTRODUCTION

Palladium is well known to show high performance for the hydrogen storage because of the small activation barrier for the surface adsorption and the exothermal reaction for the inner absorption. It has been established that the H atoms are absorbed into the octahedral vacancies of Pd fcc structure and can be absorbed to Pd/H = 1.0 in the high pressure limit owing to the result that number of the octahedral vacancies is the same with that of Pd atoms. Absorption isotherms for Pd-H system show that hydrogen absorption realized in the wide ranges of the hydrogen pressure and the sample temperature [1]. However, there are still open questions about the dynamic absorption process because most of dynamic studies carried out from the viewpoint of pressure change of surrounding hydrogen gases. Direct observation of the dynamical structural and electronic changes of Pd metal for hydrogen storage reaction, which consists of surface adsorption and following inner penetration processes [2], has not been achieved yet.

Recently, many studies have pointed out that Pd metal fine particles show the unique properties for the hydrogen storage which differs from bulk Pd [3-6]. The low-concentrate phase of Pd-H system creates the solid-solution with high solubility (α phase). In the high-concentrate phase, Pd-H chemical bondings are generated and metal hydride phase with largely expanded lattice is established (β phase). Although there is a significant phase boundary between the α phase and the β phase in the bulk Pd, the Pd metal fine particles show smooth change between two phases. For the explanation of the peculiar properties of hydrogen storage in the Pd metal fine particles, surface effect [7] and size effect [8] were suggested. It is necessary to observe the structural distortion of the Pd nanoparticles during hydrogen absorption in order to judge the formation of hydride phase over the whole of particles.

In this study, we describe the dynamic structural and electronic changes of Pd metal nanoparticles during hydrogen storage reaction with x-ray absorption fine structure (XAFS) spectra. About the XAFS observation technique, we adopt the dispersive mode. The dispersive mode which consists of the curved polychromator and the space-resolved detector enables us to observe the x-ray absorption spectra without mechanical movement at all, resulting the relatively stable and fast determination of structural parameters [9,10]. These advantages of the time resolved XAFS technique lead to new understandings about the hydrogen storage process in Pd metal nanoparitcles.

EXPERIMENT

All the Pd K-edge XAFS spectra were measured by the dispersive mode at the bending magnet beamline BL14B1 of SPring-8 [11,12]. The schematic picture is displayed in figure 1. Dispersed x-rays were obtained by a curved silicon crystal and the Si(422) reflection plane was used with the Laue configuration. Although this reflection plane contains second higher harmonics, larger intensity of light than that from all-odd reflection plane can be used. The polychromator was bent horizontally by setting it in a curved copper block cooled by the water. From the curvature with a radius of 2000 mm, x-rays with an energy range over 800 eV were obtained. $Gd_2O_2S(Tb)$ of 40 μm thickness was exposed to dispersed x-rays from the sample and emitted lights were collected using a charge coupled device (CCD) camera (640 × 480 channels, 12 bits). The intensities in the vertical direction (about 200 channels) were summed up to produce a one-dimensional spectroscopy. The horizontal focus size of the x-rays was measured to be 0.1 mm in full width at half maximum (FWHM) and the vertical size is equal to the sample pellet height for accumulating the intensity of transmitted x-rays.

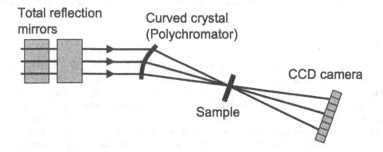

Figure 1. Schematic picture of top view of dispersive XAFS system at SPring-8, BL14B1.

Powdered γ-Al_2O_3 was used for the impregnation method with dilute aqueous palladium nitric acid, $Pd(NO_3)_2$. Following drying and calcination at 500 °C, Pd(4 wt%)/Al_2O_3 sample was prepared. The samples were reduced by hydrogen in the sample cell for the XAFS observation at 200 °C. Before the x-ray observation, the sample cell was evacuated over 30 minutes at room temperature in order to ensure the complete pure metal phase. After closing the valve connected to the vacuum line, hydrogen gas was stored in another inlet line. An opening signal from the valve connected to the inlet line was used to start the XAFS measurement. Injected hydrogen pressure was 200 kPa at sample position.

The XAFS measurements of Pd nanoparticles on Al_2O_3 during H_2 dosing were operated at room temperature by 50 Hz rate (sampling time for one spectrum is 20 ms) with the real-time-resolved mode. No data accumulation by the repetition of the reaction was adopted. Although the XAFS technique does not clearly detect the contribution of the electron scattering from a light H atom, it has been reported that a H atom affects the photoelectron wave in magnitude and phase when it is located between an absorber and a backscatter [13]. Moreover in this study, the system of hydrogen storage of Pd nanoparticles, absorbed H atoms in the Pd lattice largely elongate the interatomic distance of Pd-Pd bondings. Therefore, we can observe the atomic structure dynamics of Pd nanoparitcles even at the high frame rate.

RESULTS AND DISCUSSION

Figure 2 shows the results of the 50 Hz observation of x-ray absorption fine structure spectra for Pd/Al₂O₃ taken with dispersive mode just after the hydrogen dosing at room temperature. It is recognized that, even in the case of 50 Hz observation mode (sampling time for one spectrum is 20 ms), the oscillations in the extended region from Pd absorption edge are clearly observed. The oscillation wavelength contracts just after the hydrogen dosing, indicating the expansion of the Pd-Pd interatomic distance is caused by the hydrogen absorption. We also observed the spectra changes in the x-ray absorption near edge structure (XANES), which is particularly indicated in the negative edge energy shift [14].

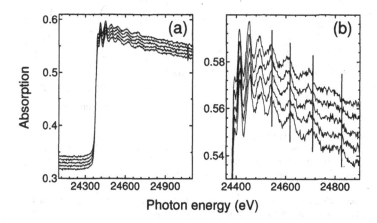

Figure 2. (a) Pd K-edge x-ray absorption fine structure spectra for Pd/Al₂O₃ just after the hydrogen dosing. Spectra were taken with dispersive mode at 50 Hz. 5 spectra just after the hydrogen dosing (during 100 ms after dosing) are depicted from bottom (first) to top (last). The spectra are vertically shifted. (b) Enlarged figure for the oscillation region of (a). Vertical lines guide eye to the contraction of the wavelength of the oscillation.

The extended x-ray absorption fine structure (EXAFS) spectra and Fourier transform intensities from the corresponding raw data of Fig. 2 are shown in Fig. 3. Contraction of the wavelength of the oscillations by the hydrogen dosing is also recognized in the EXAFS functions depicted in Fig. 3 (a). Expansion of the interatomic distance of Pd-Pd nearest neighboring configuration is clearly indicated in the Fourier transform intensities in Fig. 3 (b). The Fourier transform intensities provide the information that no distinct residues of the Pd oxide state exist.

We cannot observe any higher configurations beyond the nearest neighboring shell in the Fourier transform intensities because the Pd atoms form metal nanoparticles. The fitted back-Fourier transform spectra are also summarized in Fig. 3 (c). Fourier and back-Fourier transforms were operated with the ranges of 2.5-12.0 Å^{-1} and 1.8-2.9 Å, respectively. Normal EXAFS curve fitting procedure was used in back-Fourier transform space by *EXAFSH* code [15] with the range of 2.8-11.7 Å^{-1}. Pd metal foil was employed as the reference sample for the curve fitting of the Pd-Pd nearest neighboring shell of the Pd metal state. Operated free parameters were coordination number (CN), interatomic distance (R), and Debye-Waller factor (C_2). Those values of parameters which are determined by the calculated spectra from FEFF 8.4 code [16] for the reference sample of Pd foil at room temperature are 12, 2.75 Å, and 0.006 Å^{-2}, respectively.

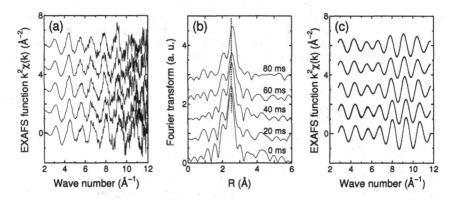

Figure 3. (a) EXAFS functions $k^2\chi(k)$ for Pd/Al_2O_3 just after the hydrogen dosing. EXAFS functions were collected with dispersive mode at 50 Hz and derived from the XAFS data set depicted in Fig. 2. (b) Fourier transform intensities of EXAFS spectra in (a). (c) Back-Fourier transform spectra of Fourier transform intensities in (b). Dotted lines show the fitted ones.

Figure 4 shows the variations of the XAFS parameters determined by the EXAFS curve fitting. The edge energy shift was determined from the XANES spectra. From the change of the interatomic distance just after the hydrogen dosing, a large expansion of the Pd-Pd interatomic distance from 2.73 to 2.83 Å is clearly understood. The expansion of the Pd particles is completed in about 50 ms after the 200 kPa H_2 dosing. It was revealed that the Pd lattice is directly changed to the hydride phase in short time. Along with the expansion of the Pd-Pd interatomic distance, negative shift of the edge energy position and decrease of the CN are also recognized. The shift of the edge position is brought about by the electron donation from hydrogen to Pd particles, or widened Pd band structure due to the formation of the Pd-H bondings. As for the case of CN, smaller value after the hydrogen dosing suggests the increase of surface area due to the dispersion of the particle. It is also noted that the Debye-Waller factor C_2 does not show the significant change during hydrogen dosing. Because the Debye-Waller factor C_2 is the second cumulant of the interatomic distance, the large shift of the value of the interatomic distance can influence the value of the Debye-Waller factor C_2. This result indicates that the similar Pd-Pd bond lengths are created in the metal hydride phase of the particles and denies the separation model of surface and inner regions.

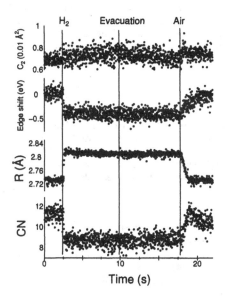

Figure 4. Variations of XAFS parameters of Pd/Al₂O₃ during gas changes at room temperature. XAFS spectra were measured by dispersive mode at 50 Hz. Applied hydrogen pressure was 200 kPa and applied air pressure was 100 kPa.

After the evacuation of the sample cell (at about 10 s in Fig. 4), any clear changes are not observed for all parameters in this time scale. We observed that the hydrogen desorption by evacuation proceeds in the time scale of 10 minutes. From the viewpoints of the hydrogen solubility in the Pd particles, strong Pd-H chemical bonding suppresses the movement of the hydrogen in the hydride phase. In another point of the desorption mechanism, the large surface potential against the vacuum side inhibits the hydrogen desorption. On the other hand, air injection at about 18 s moves all the parameters back to the initial values in about 1 second. The chemical reaction between absorbed hydrogen and the oxygen included in the air causes the disappearance of the all absorbed hydrogen from Pd metal nanoparticles in the short time scale. Judging from the comparison between the evacuation and the air injection treatments, it is revealed that the surface desorption energy barrier is the key issue for the hydrogen desorption process.

CONCLUSIONS

Dynamic local structural change of Pd nanoparticles on alumina surface during hydrogen absorption was directly observed by x-ray absorption fine structure spectroscopy with dispersive mode. In situ and real-time-resolved observation enabled us to determine main four parameters of x-ray absorption spectroscopy even in the case of 50 Hz study. Pd nanoparticles directly change to the hydride phase in 50 ms at 200 kPa of hydrogen pressure. Almost no structural distortion was observed after the lattice expansion, which indicates that all Pd-Pd bondings have similar elongated bond length in the hydride phase. This is the first direct observation of the dynamic structure change of the Pd nanoparticles with the high frame rate of 50 Hz.

ACKNOWLEDGMENTS

We are grateful to Dr. A. Suzuki and Professors Y. Inada and M. Nomura of Institute of Materials Structure Science for stimulating discussions. We also thank Mr. K. Kato, Dr. T. Uruga and Dr. H. Tanida for constructing the dispersive XAFS system. This work was supported by the New Energy and Industrial Technology Development Organization (NEDO) under "Advanced Fundamental Research Project on Hydrogen Storage Materials".

REFERENCES

1. H. Frieske and E. Wicke, *Ber. Bunseges. Physik. Chem.* **77**, 50 (1973).
2. B. D. Kay, C. H. F. Peden and D. W. Goodman, *Phys. Rev. B* **34**, 816 (1986).
3. M. Yamauchi, R. Ikeda, H. Kitagawa and M. Takata, *J. Phys. Chem. C* **112**, 3294 (2008).
4. M. Wilde and K. Fukutani, *Phys. Rev. B* **78**, 115411 (2008).
5. M. Wilde, K. Fukutani, M. Naschitzki and H. -J. Freund, *Phys. Rev. B* **77**, 113412 (2008).
6. H. Kobayashi, M. Yamauchi, H. Kitagawa, Y. Kubota, K. Kato and M. Takata, *J. Am. Chem. Soc.* **130**, 1818 (2008).
7. C. Sachs, A. Pundt, R. Kirchheim, M. Winter, M. T. Reets and D. Fritsh, *Phys. Rev. B* **64**, 075408 (2001).
8. J. A. Eastman, L. J. Thompson and B. J. Kestel, *Phys. Rev. B* **48**, 84 (1993).
9. T. Matsushita and R. P. Phizackerley, *Jpn. J. Appl. Phys.* **20**, 2223 (1981).
10. U. Kaminaga, T. Matsushita and K. Kohra, *Jpn. J. Appl. Phys.* **20**, 355 (1981).
11. Y. Okajima, D. Matsumura, Y. Nishihata, H. Konishi and J. Mizuki, *AIP Conference Proceedings* **879**, 1234 (2007).
12. D. Matsumura, Y. Okajima, Y. Nishihata, J. Mizuki, M. Taniguchi, M. Uenishi and H. Tanaka, *J. Phys.: Conf. Ser.* **190**, 012154 (2009).
13. B. Lengeler, *Phys. Rev. Lett.* **53**, 74 (1984).
14. R. J. Davis, S. M. Landry, J. A. Horsley and M. Boudart, *Phys. Rev. B* **39**, 10580 (1989).
15. T. Yokoyama, H. Hamamatsu and T. Ohta, *EXAFSH*, The Unversity of Tokyo, 2nd Ed. (1998).
16. A. L. Ankudinov, A. I. Nesvizhskii and J. J. Rehr, *Phys. Rev. B* **67**, 115120 (2003).

Batteries and Photosynthesis

Mater. Res. Soc. Symp. Proc. Vol. 1262 © 2010 Materials Research Society 1262-W07-10

X-ray Detection on Fe-H Vibrations

Hongxin Wang,[1,2] Yisong Guo,[1] Saeed Kamali,[1] and Stephen P. Cramer[1,2]

[1] Department of Applied Science, University of California, Davis, CA 95616, USA
[2] Lawrence Berkeley National Lab, Berkeley, CA 94720, USA

ABSTRACT

X-ray detection on hydrogen related events is difficult due to its extremely small scattering factor. In this report, we have used nuclear resonance vibrational spectroscopy (NRVS) to examine the nature of the Fe–H vibrational modes in several FeH model complexes, which shines light on the possible measurements on the Fe-H vibrations inside real biological systems in the future.

INTRODUCTION

Hydrogenase catalyzes the reversible dihydrogen (H_2) production and oxidation [1,2], and is one of the most important enzymes in nature. The enzyme and/or its functional models are possible candidates for bio-hydrogen production as clean fuel in the future [3].

For NiFe hydrogenases, Ni-A (as-isolated state) has a oxo or hydroxo ligand bridging in between the Ni and the Fe inside the NiFe center, while Ni-R is proposed to have a hydride in the same position instead [1].

In the fields of energy and materials science, iron, iron oxide and other iron complexes are important materials used for solar hydrogen production [4] and hydrogen storage [5]. For these materials and their applications, the interaction between Fe and H is highly interested as well.

While X-ray spectroscopy is good for its element specificity, its detection on light elements, especially on hydrogen, is usually very difficult. Therefore no direct X-ray observation has been reported so far for the Fe-H interactions mentioned above.

Nuclear resonance vibrational spectroscopy (NRVS) measures the vibrational modes coupled with the ^{57}Fe nuclear transition at 14.4 KeV. It is an element and isotope specific measurement. In this report, we will illustrate how to use the method to probe the Fe-H vibrational modes in various FeH model complexes, which will serve as the first step towards the Fe-H measurements for real biological systems and/or other complicated FeH systems in the future.

EXPERIMENTS

The NRVS spectra were recorded with a published procedure [6,7] at APS (ID03) in the USA and at SPring-8 (BL09XU) in Japan. At APS, as shown in figure 1, the X-ray beam from the synchrotron radiation storage ring went through a high heat load monochromator (pre-mono, $C(1,1,1)xC(1,1,1)$) to produce 14.4 KeV radiation with 1 eV in energy resolution and then a high-resolution monochromator (HRM, $2Si(10,6,4)x2Si(4,0,0)$) to produce 14.4 KeV radiation with 1 meV (8 cm^{-1}) resolution. At SPring-8, the pre-mono ($Si(1,1,1)xSi(1,1,1)$) produced 14.4KeV radiation with 1.5 eV resolution, while and the HRM ($Ge(4,2,2)x2Si(9,7,5)$) produced 14.4KeV radiation with 1.2 meV resolution.

During all NRVS measurements, the samples were maintained at a low temperature using a liquid helium cryostat. The real sample temperatures were calculated [6] using the ratio of anti-stokes to stokes intensities according to $S(-E)/S(E)=\exp(-E/kT)$ [6]. The temperatures were usually below 100K.

Delayed nuclear fluorescence and Fe K fluorescence (from internal conversion) were recorded with a single avalanche 1 cm^2 square APD at the APS and with a four-element APD array at SPring-8. NRVS spectra were recorded with a step size of 0.25 meV, and each point was the sum of 5 sec.

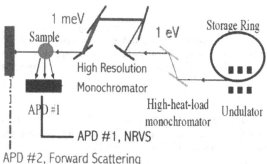

Figure 1. Setup concept for a NRVS experiment at APS ID03.

Partial vibrational density of state (PVDOS) was calculated from the raw NRVS spectra using PHOENIX program [6,7]. In the following text, NRVS spectra mean the PVDOS, not the raw NRVS spectra.

NRVS spectrum from FeCl$_4$[NEt$_4$] was used to calibrate the beam energy to a reported value (the peak centroid) of 47.1 meV (or 380 cm^{-1}) [6].

RESULTS AND DISCUSSIONS
The results of the NRVS studies of Fe-H vibrations are summarized as below, beginning with the iron complex with six hydride ligands (H$^-$), and going to more dilute (less H$^-$) iron complexes.

3.1 [FeH$_6$]$_4^-$
This NRVS spectrum was measured previously in this group as a feasibility study, with an eye on possible hydrogenase projects in the future [8].

We compared the isotopomers of [FeH$_6$]$_4^-$ vs. [FeD$_6$]$_4^-$ (D stands for deuteride) to see if Fe-H(D) stretches, which involve very little Fe motion, would be visible by NRVS. As illustrated in figure 2, Fe-D stretching mode was visible at 1122 cm^{-1}, while the Fe-H stretch mode at 1569 cm^{-1} (according to IR) was not observed [8]. We also noticed the surprising strength of the H–Fe–H and D–Fe–D bending modes at 787 cm^{-1} and 571 cm^{-1} respectively.

3.2 FeHCO and FeHN₂

In figure 2, we have illustrated the feasibility of probing Fe-H(D) vibrational modes inside [FeH(D)₆]⁻ from a previous experiment. However, the iron will not have six hydride ligands in real hydrogenase enzymes and/or many model complexes. For example, in a NiFe hydrogenase, one H⁻ is proposed to bridge the Ni and the Fe (X position in Ni-R, figure 3a) [1]. To further reveal the situation closer to the reality, we selected two FeH model complexes - each has only one H⁻ ligand to Fe center, as shown in figure 3 (b and c).

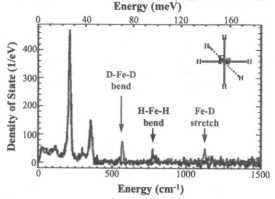

Figure 2. NRVS Spectra of $[FeH_6]_4^-$ and $[FeD_6]_4^-$

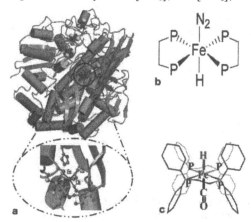

Figure 3. Structures of NiFe center from *D. gigas* hydrogenase in Ni-A state (a), Fe(H)(N₂)(DMeOPrPE)₂ (b), and Fe(H)(CO)((PPh₂)₄ (c). The X position (orange color) in (a) is the possible hydride position in Ni-R.

As shown in figure 4, these two iron complexes which have single-hydride ligand also showed significant NRVS signals in the X-Fe-H (730-750 cm⁻¹) and X-Fe-D (600-610 cm⁻¹)

bending regions, where X=C or N in the diatomic CO or N_2 ligands in the *trans-* position. The isotopic shifts in between H and D were as expected - less than √2 for both complexes. The P-Fe-H modes seem more spread and couple with other modes in the lower energy regions [9].

For the FeHCO and FeDCO complexes, the peaks at 540-560 cm⁻¹ region were contributed from Fe-CO stretching and Fe-C-O bending vibrations [9] slightly coupled with C-Fe-H(D) bending vibrations. The intensity at ~740 cm⁻¹ in DFeCO was probably due to the H impurity when the complex was not 100% labeled with D.

Figure 4 has clearly shown: 1) single Fe-H(D) sites can be unambiguously observed by NRVS; 2) X-Fe-H(D) (X=C, N) bending modes do not overlap with the biological relevant vibrational modes of Fe-C, Fe-N, Fe-P, which exist in these two model complexes as well. According to previous studies, Fe-S is in an even lower energy region of 100-400 cm⁻¹ [7,10]. Thus X-Fe-H(D) bending modes observed at 580-800 cm⁻¹ can be used as bench marks for tracking Fe-H(D) sites inside various iron complexes.

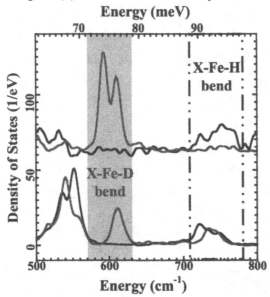

Figure 4. NRVS Spectra of Fe(H/D)(N$_2$)(DMeOPrPE)$_2$ (top) and Fe(H/D)(CO)(PPh$_2$)$_4$ (bottom). The black and red lines are for FeH / FeD complexes respectively.

3.3 NRVS Advantages and Perspectives in Enzymes

NRVS measures the vibrational modes coupled with the ^{57}Fe nuclear transition at 14.4 KeV. It is element and isotope (^{57}Fe) specific; the transitions are allowed as long as the vibrational modes involves a ^{57}Fe motion in the direction of incoming X-ray; both spectral intensities and frequencies can be quantitatively simulated with a normal mode analysis [6,7].

These advantages enable us to: 1) easily distinguish Fe-H(D) modes from other non-iron vibrations in the same region and; 2) accurately identify Fe-H(D) vibrations by frequency-intensity dual simulation of the spectra. Therefore NRVS is best for measuring Fe-H(D) inside complicated systems, such as hydrogenase.

However, NRVS application on a real biological system could be complicated by the several issues as below:

1) diluteness of the iron centers within the protein environment could make the NRVS signal and signal-to-noise level less favorable;

2) greater complexity afforded by a decrease in symmetry could make the peak positions and NRVS intensity more spread; and

3) Fe-CO or some other Fe-C vibrational modes over 600 cm^{-1} could overlap with the Fe-D bending mode(s).

Nevertheless, the evidence presented here clearly demonstrates the feasibility to probe X-Fe-H(D) related bending modes in the region of 580-800 cm^{-1}. Therefore NRVS is surely applicable as a method for detecting these Fe-H(D) vibrations and for identifying Fe-H(D) sites.

SUMMARY
The results presented in this report illustrate that nuclear resonance vibrational spectroscopy (NRVS) can have a strong role in determining iron hydride / iron-deuteride bonding in Fe-H(D) complexes. With the development of the synchrotron radiation facilities, nuclear spectroscopy beamlines as well as the detection technologies, this element and isotope specific spectroscopy will become more and more readily available for studying real biological molecules concerning Fe-H sites.

ACKNOWLEDGEMENTS
We thank Drs E. Alp, J. Zhao and W. Sturhahn at the APS and Y. Yoda at SPring-8 (Proposal No.=2009A0015) for assistance with NRVS measurements; T. Rauchfuss at UIUC, and D. Tyler at U. Oregon for the samples. This work was funded by grants (to SPC): NSF (CHE-07453535), NIH (GM-44380, GM-65440), and DOE-OBER. The APS is supported by the DOE and SPring-8 is supported by JASRI.

REFERENCES
1   W. Lubitz, E. Reijerse and and M. van Gastel. *Chem. Rev.* **107**, 4331 (2007).
2   H. Wang, C. Ralston, D. Patil, R. Jones, W. Gu, M. Verhagen, M. Adams, P. Ge etc., and S. P. Cramer, *J. Am. Chem. Soc*, **122**, 10544 (2000).
3   U.S. Department of Energy Office of Science - Biohydrogen Production (October 28, 2009), from website: http://genomicsgtl.energy.gov/benefits/biohydrogen.shtml
4   P. Charvin, S. Abanades, F. Lemort, and G. Flamant, *Energy & Fuels* **21**, 2919 (2007).
5   Z. Huanga, Z. Guoa, A. Calkab, D. Wexlerb, C. Lukeyc and H. Liua, *J. Alloys & Compounds* **422**, 299 (2006).

6   M. Smith, Y. Xiao, H. Wang, S. George, D. Coucouvanis, M. Koutmos, W. Sturhahn, E. Alp, J. Zhao, S. Cramer, *Inorg. Chem.*, **44**, 5562 (2005).

7   Y. Xiao, H. Wang, S. George, M. Smith, M. Adams, F. Jenney, W. Sturhahn, E. Alp, J. Zhao, Y. Yoda, A. Dey, E. Solomon, S. Cramer, *J. Am. Chem. Soc.*, **128**, 7608 (2006)

8   U. Bergmann, W. Sturhahn, D. Linn, F. Jenney, M. Adams, K. Rupnik, B. Hales, S. Cramer, *J. Am. Chem. Soc,* **125**, 4016 (2003).

9   V. Pelmenschikov, Y. Guo, H. Wang, S. P. Cramer, D. A. Case, *Faraday Discuss.* Accepted (2010).

10   S. Cramer, Y. Xiao, H. Wang, Y. Guo, M. Smith, *Hyperfine Interact* **170**, 47 (2006).

Solar Cells

Mater. Res. Soc. Symp. Proc. Vol. 1262 © 2010 Materials Research Society   1262-W08-03

XANES studies on Eu-doped fluorozirconate based glass ceramics

Bastian Henke[1,2], Patrick Keil[3], Christian Paßlick[2], Dirk Vogel[3], Michael Rohwerder[3], Marie-Christin Wiegand[4], Jacqueline A. Johnson[5], and Stefan Schweizer[1,2]

[1]Fraunhofer Center for Silicon Photovoltaics, Walter-Hülse-Str. 1, 06120 Halle (Saale), Germany
[2]Centre for Innovation Competence SiLi-nano®, Martin Luther University of Halle-Wittenberg, Karl-Freiherr-von-Fritsch-Str. 3, 06120 Halle (Saale), Germany
[3]Department of Interface Chemistry and Surface Engineering, Max-Planck-Institut für Eisenforschung GmbH, Max-Planck-Str. 1, 40237 Düsseldorf, Germany
[4]Department of Physics, University of Paderborn, Warburger Str. 100, 33100 Paderborn, Germany
[5]Department of Materials Science and Engineering, University of Tennessee Space Institute, Tullahoma, TN 37388, USA

ABSTRACT

The influence of adding InF_3 as a reducing agent on the oxidation state of Eu in fluoro-chloro- (FCZ) and fluorobromozirconate (FBZ) glass ceramics was investigated using x-ray absorption near edge (XANES) and photoluminescence (PL) spectroscopy. For both materials, it was found that InF_3 decreases the Eu^{2+}-to-Eu^{3+} ratio significantly. PL spectroscopy proved that an annealing step leads to the formation of Eu-doped $BaCl_2$ and $BaBr_2$ nanocrystals in the FCZ and FBZ glasses, respectively. In the case of FCZ glass ceramics the hexagonal phase of $BaCl_2$ could be detected in indium-free and InF_3-doped ceramics, but only for InF_3 containing FCZ glass ceramics a phase transition of the nanoparticles from hexagonal to orthorhombic structure is observed. For the FBZ glass ceramics, the hexagonal phase of $BaBr_2$ can be formed with and without indium doping, but only in the indium-free case a phase transition to orthorhombic $BaBr_2$ could be found.

INTRODUCTION

A class of fluorochloro- (FCZ) and fluorobromozirconate (FBZ) glass ceramics has been developed for ionizing radiation detection. The composition of these materials is derived from a standard ZBLAN glass, i.e. a fluoride glass, made from Zr, Ba, La, Al, and Na fluorides. Several synthesis routes have been derived to maintain Zr in the 4+ state where the addition of a small amount of InF_3 plays a key role in the glass production process. The glasses under investigation were additionally doped with Cl or Br ions by partial substitution of the fluorides BaF_2 and NaF for BaX_2 and NaX (X = Cl, Br), respectively. This enabled precipitation of BaX_2 nanocrystals within the glass upon appropriate annealing.

For ionizing radiation detection, Eu is added to the glass for optical activation. Depending on the structural phase of the BaX_2, the glass-ceramic material can act either as a scintillator (able to convert ionizing radiation to visible light), or as a storage phosphor (able to convert the radiation into stable electron-hole pairs, which can be released at a later time with a scanning laser beam).

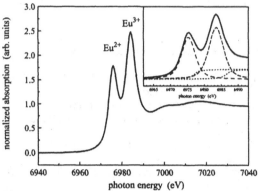

Figure 1. Normalized XANES spectra of the as-made, indium-free, Eu-doped FBZ glass. A pseudo-Voigt function (dashed curve) and an arctan function (dotted curve) were used to fit each of the characteristic WLs of Eu^{2+} and Eu^{3+} and the absorption edge, respectively (see inset).

To obtain luminescence from these materials, the critical component is the Eu^{2+}, in particular the Eu^{2+} present in the BaX_2 nanocrystals. However, there is always a significant amount of Eu^{3+} in the glass which reduces the performance of these materials as both scintillators and storage phosphors. X-ray absorption near edge spectroscopy (XANES) measurements showed that one source was found to be in the as-received EuF_2, in which Eu^{2+} is oxidized to Eu^{3+} before and/or during glass melting [1]. In such a complex system, the addition of each new chemical and each change in processing affects the material properties. In an earlier work on Eu-doped FCZ glass ceramics, the influence of InF_3 and remelting on the Eu^{2+}-to-Eu^{3+} ratio was investigated [2]. It was shown that the addition of InF_3 decreases the Eu^{2+}-to-Eu^{3+} ratio whereas remelting increases the Eu^{2+}-to-Eu^{3+} ratio.

This work is a step on the road to identify the role of each chemical component and to improve the material for luminescence applications such as ionizing radiation detection. In particular, we investigate the influence of adding InF_3 on the Eu^{2+}-to-Eu^{3+} ratio in FBZ glass ceramics and compare the results with those from FCZ glass ceramics.

EXPERIMENT

XANES

Fig. 1 shows the normalized Eu L_3 XANES spectrum of the as-made Eu-doped FBZ glass without InF_3 doping. The two resonances at about 6976 eV and 6986 eV are the characteristic white lines (WLs) of Eu^{2+} and Eu^{3+} [3] and are associated with a dipole-allowed transition from a $2p_{3/2}$ core level into an empty $5d$ state. The Eu^{2+}-to-Eu^{3+} ratio can be determined by building the relative intensity of the two WLs [4, 5]. A pseudo-Voigt function and an arctan function were used to fit each WL and the absorption edge, respectively. The corresponding fitted curve (solid curve) and the individual fitted components (dashed and dotted curves) are shown in the inset of

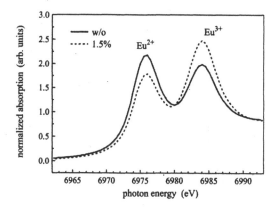

Figure 2. Normalized XANES spectra of the as-made Eu-doped FBZ glasses without (solid curve) and with 1.5 mol% InF₃ (dotted curve).

Fig. 1. The mole ratio M_{Eu} of Eu^{2+} to Eu^{3+} can be determined from the area ratio A_{Eu} of the WLs $M_{Eu} = RA_{Eu}$, where R is a constant that is related to the $2p \rightarrow 5d$ transition probabilities of Eu^{2+} and Eu^{3+} [5]. Due to the low Eu-doping level of 3 mol%, the R value is set to 1.5, which was determined for dilute systems from an "in situ" XANES analysis of electrochemical conversion between Eu^{2+} and Eu^{3+} in a $EuCl_3$ solution [5]. For the Eu-doped FCZ glasses the curve fitting results and the calculated mole fractions are listed in [2].

Fig. 2 shows the normalized Eu L_3 XANES spectra of as-made FBZ glass ceramics without (solid curve) and with 1.5 mol% (dashed curve) InF₃ doping. While there is no significant shift in the energy of the XANES WLs in any of the spectra, the WL intensity and thus the Eu^{2+}-to-Eu^{3+} mole ratio changes significantly upon increasing the InF₃ doping level. In Fig. 3 the effect of thermal annealing on the Eu^{2+}-to-Eu^{3+} ratio of the Eu-doped FCZ and FBZ glass ceramics is presented. Obviously, thermal annealing has no significant influence on the Eu^{2+}-to-Eu^{3+} ratio. The InF₃ doping, however, changes the Eu^{2+}-to-Eu^{3+} ratio significantly. Thus, the InF₃ doping leads generally to a decrease in the Eu^{2+}-to-Eu^{3+} ratio from 2.2 (0 mol% InF₃) to 1.2 (1.5 mol% InF₃) and from 1.8 (0 mol% InF₃) to 0.85 (0.5 mol% InF₃) for the FBZ and FCZ glasses, respectively.

Photoluminescence (PL)

PL spectra of differently annealed Eu-doped FCZ and FBZ samples (with and without InF₃) are plotted in Fig. 4. The indium-free FCZ sample annealed at 270 °C (top left, dotted curve) shows the typical PL spectrum for a glass ceramic containing predominantly hexagonal phase $BaCl_2$ particles [6]. The peak at 406 nm is attributed to the $5d$-$4f$ transition of Eu^{2+} in hexagonal $BaCl_2$; the origin of the less intense but broader emission at 485 nm is unknown. When annealing up to 290 °C (solid curve), the intensity of the 485 nm band lessens but there is no phase transformation from hexagonal to orthorhombic phase $BaCl_2$, as was reported in [6]. The

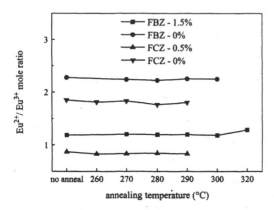

Figure 3. Eu^{2+}-to-Eu^{3+} mole ratio vs. annealing temperature.

InF$_3$ doped FCZ sample (top right) also shows the hexagonal phase after annealing at 270 °C, but annealing at 290 °C leads to the formation of orthorhombic phase BaCl$_2$: The 406 nm band is shifted to 401 nm while the 485 nm band is almost gone.

In contrast to the FCZ samples, the Eu-doped FBZ samples without indium (bottom left) show a phase transformation to orthorhombic BaBr$_2$ upon increasing the annealing temperature from 280 to 300 °C. Here, the $5d$-$4f$ transition of Eu^{2+} in hexagonal BaBr$_2$ peaks at 407 nm and is shifted to 402 nm for orthorhombic BaBr$_2$ [7]. In analogy to the FCZ samples, the 485 nm band is only observed in glass ceramics containing hexagonal phase BaBr$_2$ nanoparticles [7]. Interestingly, the FBZ samples doped with InF$_3$ (bottom right) do not show a phase transition, but only hexagonal phase BaBr$_2$ particles.

DISCUSSION

The results show that the Eu^{2+}-to-Eu^{3+} ratio in Eu-doped FCZ and FBZ glasses depends significantly on the addition of InF$_3$. Without the InF$_3$ additive, a higher fraction of Eu^{2+} is produced. However, the quality of the glasses also depends on the addition of indium and the melting process. Samples with lower amounts of InF$_3$ show a large number of black precipitates on the surface. These black specks in the glass ceramics arise from Zr^{3+}, which is reduced from Zr^{4+} [8]. This problem is alleviated in the presence of InF$_3$ as InF$_3$ acts as a mild oxidizer. If any Zr^{3+} does form, the In^{3+} oxidizes it back to Zr^{4+}, and in this process indium is reduced to In^{1+}. It appears that while InF$_3$ is added as an oxidizer to maintain Zr in the 4+ state, it also oxidizes some of the Eu^{2+} to Eu^{3+}.

In all annealed samples, crystallization of hexagonal phase BaCl$_2$ and BaBr$_2$, respectively, can be found. The indium additive, however, affects the crystallization properties of the FCZ and FBZ glasses: Analogous to previous work [6], for the InF$_3$-containing FCZ glass ceramics a phase transformation from hexagonal to orthorhombic phase BaCl$_2$ is observed. Here, indium is required to produce the desirable orthorhombic phase for the material to work as a

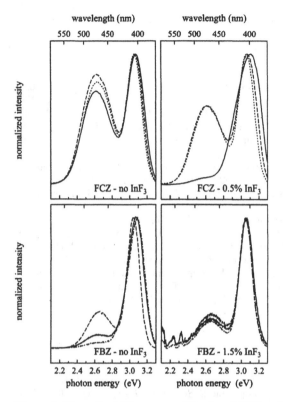

Figure 4. Normalized PL spectra of Eu-doped FCZ and FBZ glass ceramics annealed at 270, 280, 290 °C and 280, 290, 300 °C, respectively, for 20 min (270 °C – dotted, 280 °C – dashed, 290 °C – solid, 300 °C – dashed-dotted). The PL was excited at 280 nm. All spectra have been normalized to their most intense emission.

storage phosphor. The hexagonal phase is formed at temperatures below 290 °C, above this temperature the orthorhombic phase is formed. Contrary to the FCZ glasses, the InF_3-containing FBZ glasses do not show a phase transformation to orthorhombic $BaBr_2$, but the indium-free FBZ glass ceramics do.

This work is a step on the road to identify the role of each chemical component and to improve the material for luminescence applications such as ionizing radiation detection. InF_3 has both advantages and disadvantages. For example, it has proved to be essential to glass quality. Without indium, the quantity of black precipitates (reduced Zr) is unacceptably high. In addition, InF_3 is also important for the phase transformation: Without the addition of In, no phase transformation from hexagonal to orthorhombic phase $BaCl_2$ nanoparticles can be found in FCZ glass ceramics, while for FBZ glass ceramics it is vice versa. Experiments are ongoing to improve the glass ceramics both for ionizing radiation detection and frequency conversion, such as a two-step preparation process and/or the addition of codopants such as divalent samarium.

ACKNOWLEDGMENTS

This work was supported by the FhG Internal Programs under Grant No. Attract 692 034. In addition, the authors would like to thank the German Science Foundation ("Deutsche Forschungsgemeinschaft") for their financial support (DFG Project No. PAK88) and the Federal Ministry for Education and Research ("Bundesministerium für Bildung und Forschung") for their financial support within the Centre for Innovation Competence SiLi-nano® (project number 03Z2HN11).

The project described was also supported by the National Institute of Biomedical Imaging and Bioengineering (NIBIB) (Grant No. 5R01EB006145-02). The content is solely the responsibility of the authors and does not necessarily represent the official views of the NIBIB or the National Institutes of Health.

Portions of this work were performed at the DuPont-Northwestern-Dow Collaborative Access Team (DND-CAT) located at Sector 5 and at the X-ray Operations and Research beamline 12-BM of the Advanced Photon Source (APS). DND-CAT is supported by E.I. DuPont de Nemours & Co., The Dow Chemical Co., and the State of Illinois. Use of the APS at the Argonne National Laboratory was supported by the U.S. Department of Energy, Office of Science, Office of Basic Energy Sciences, under Contract No. DE-AC02-06CH11357. The authors would like to thank Qing Ma of the DND-CAT and Nadia Leyarovska for their help with the experiment.

REFERENCES

1. G. Chen, J. A. Johnson, J. Woodford, and S. Schweizer, *Appl. Phys. Lett.* **88**, 191915 (2006).
2. B. Henke, C. Paßlick, P. Keil, J. A. Johnson, and S. Schweizer, *J. Appl. Phys.* **106**, 113501 (2009).
3. J. Rohler, in "Handbook on the Physics and Chemistry of Rare Earths", edited by K. A. Gschneidner, Jr., L. Eyring, and S. Hufner (North-Holland, Amsterdam, 1987), vol. 10, pp. 453-545.
4. Y. Takahashi, G. R. Kolonin, G. P. Shironosova, I. I. Kupriyanova, T. Uruga, and H. Shimizu, *Miner. Mag.* **69**, 179 (2005).
5. M. R. Antonio, L. Soderholm, and I. Song, *J. Appl. Electrochem.* **27**, 784 (1997).
6. S. Schweizer, L. W. Hobbs, M. Secu, J.-M. Spaeth, A. Edgar, G. V. M. Williams, *J. Appl. Phys.* **97**, 083522 (2005).
7. A. Edgar, M. Secu, G. V. M. Williams, S. Schweizer, and J.-M. Spaeth, *J. Phys.: Condens. Matter* **13**, 6259-6269 (2001).
8. J. M. Parker, *Annu. Rev. Mater. Sci.* **19**, 21 (1989).

Porous Media and Disordered Systems

Mater. Res. Soc. Symp. Proc. Vol. 1262 © 2010 Materials Research Society   1262-W10-05

Correlating Small Angle Scattering Spectra to Electrical Resistivity Changes in a Nickel-Base Superalloy

Ricky L. Whelchel[1], V. S. K. G. Kelekanjeri[1], Rosario A. Gerhardt[1,*], Jan Ilavsky[2], and Ken C. Littrell[3]

[1]School of Materials Science and Engineering, Georgia Institute of Technology, Atlanta, GA, 30332, U.S.A.

[2]X-ray Operations Division, Argonne National Laboratory 9700S Cass Ave, Bldg. 438E, Argonne, IL 60439, U.S.A.

[3]High Flux Isotope Reactor, Oak Ridge National Laboratory, Oak Ridge, TN 37831, U.S.A.

*email: rosario.gerhardt@mse.gatech.edu

ABSTRACT

Waspaloy specimens aged at 800°C from 0.5h to 88.5h were evaluated via small angle neutron scattering (SANS), ultra small angle X-ray scattering (USAXS), electrical resistivity, and SEM. The average γ' precipitate size and volume fraction, obtained from modeling the small angle scattering data, was used to calculate a figure of merit of electron scattering. This figure of merit is designed to correlate the electron scattering ability of the material to the precipitate microstructure. The USAXS data shows a secondary precipitate population at smaller diameters that is absent from the SANS data, since the SANS measurements were not obtained at high enough values of Q. It is believed that this secondary population makes the USAXS-derived figure of merit more sensitive to the actual measured resistivity response than the SANS-derived values; however, the SANS derived primary precipitate sizes are believed to more accurate due to a larger sample volume.

INTRODUCTION

Waspaloy is a nickel-base superalloy used primarily in disc rotors in gas turbine engines. Waspaloy is ideal for this purpose due to the material's increased high temperature strength and creep resistance due to the formation of nanometer-scale precipitate phases (γ') within a solutionized matrix phase (γ) via heat treatment. The pre-existing precipitate phases formed via heat treatment can evolve with in-service thermal exposure in gas turbine engines, resulting in evolving mechanical properties of the bulk superalloy components. Consequently, it is desirable to monitor this evolution in mechanical properties non-destructively.

Electrical resistivity is one method by which the precipitate microstructural evolution may be sensed. Electrical resistivity is sensitive to the formation of small phases and the removal of precipitate phase solute [1]. Small precipitates, on the order of 1nm, have significant electron scattering ability and are the dominant features affecting electrical resistivity changes at early stages of the aging process. Solute removal due to precipitation results in a more pure, and thus more conductive, matrix phase. The solute removal process becomes dominant at later stages of aging, when large volume fractions of precipitates are formed; consequently a large amount of precipitate phase solute is removed at this stage of the aging process [1].

In order to use the electrical resistivity technique to monitor the γ' microstructure, it is necessary to correlate the changes in the resistivity response to the changing precipitate microstructure with aging. Small angle scattering (SAS) techniques provide a means to obtain precipitate microstructural information in Waspaloy throughout the sample volume [2, 3].

In this paper, ultra small angle X-ray scattering (USAXS) and small angle neutron scattering (SANS) measurements are compared for Waspaloy samples aged at 800°C. The microstructural information obtained from the SAS spectra was used to calculate a figure of merit of electron scattering described elsewhere [2, 3]. This figure of merit is designed to model the ability of the precipitate microstructure to scatter electrons, and thus promote changes to the material's electrical resistivity. The figure of merit is then compared to measured electrical resistivity data.

EXPERIMENT

Specimen preparation

Industrial grade Waspaloy bars were obtained from Fry Steel with an approximate composition, ignoring trace elements, as follows (in at%: Ni 56.1, Cr 21.2, Co 12.3, Ti 3.6, Al 2.7, Mo 2.5 and Fe 1.3). The bars were given a solution treatment at 1145°C for 4 hours followed by quenching in an ~5wt% brine solution. The specimens were subsequently aged at 800°C for times ranging from 0.5 to 88.5h with intermediate quenching and sampling taking place after each aging step. SANS and resistivity specimens were sectioned to be approximately 2mm thick, followed by polishing with a series of 400, 800, and 1200 US grit grinding papers. Microscopy and USAXS specimens were sectioned to be 1mm thick. The microscopy specimens were initially polished with 9μm and 3μm diamond suspensions. Final polishing consisted of a suspension of four parts of 0.05μm colloidal silica plus one part of 30% H_2O_2, followed by a suspension of 0.05μm colloidal alumina. A preferential γ' etchant was utilized to observe γ-γ' contrast, and a LEO 1530 scanning electron microscope was utilized for both imaging and energy dispersive spectroscopy (EDS) measurements. USAXS specimens were further ground with a series of 400, 800, and 1200 US grit grinding papers via a Gatan disc grinder until final specimen thicknesses of approximately 10 to 20μm were achieved. Specimens with this small thickness were necessary due to high absorption of X-rays in the alloy.

Small angle scattering

USAXS measurements were performed at the 32 ID beamline at Argonne National Laboratory's Advanced Photon Source using 11.9 keV X-rays for an approximate Q range from 10^{-4} to 1 Å^{-1}. SANS measurements were performed at the CG-2 beamline at Oak Ridge National Laboratory's High Flux Isotope Reactor for an approximate Q range from 10^{-3} to 0.1 Å^{-1}. All data modeling was conducted through programs created via the Igor Pro software from Wavemetrics, Inc. The USAXS data was reduced via the Indra 2 package [4], and the SANS data was reduced via the SPICE SANS reduction package [5]. All data modeling was conducted via the Irena 2 package [6]. Both the USAXS and SANS data were modeled with spherical scattering objects, as the γ' precipitates are known to be spherical in Waspaloy. The γ- γ' scattering contrast was calculated from EDS data on the etched microscopy specimens.

Electrical resistivity testing

DC four-point probe resistivity measurements were obtained with a Signatone model S301-6 probe station attached to a Signatone SP4-40045TFS four probe head. Current was applied using the delta mode setting of a Keithley 6221 AC/DC current source, and voltage measurements

146

were obtained using a Keithley 2182A nanovoltmeter. Resistance measurements were converted to resistivity via geometric correction factors developed by Kelekanjeri and Gerhardt [7].

Figure of merit modeling

The figure of merit of electron scattering, designated as η', is given by equation 1, and is described in detail elsewhere [2, 3]. The purpose of this figure of merit is to relate microstructural information, such as volume fraction (f_v) and average precipitate radius ($<r>$) to the expected electrical resistivity response.

$$\eta' = \frac{\eta_g + \eta_c}{|\eta_{max}|} = \frac{\sum_i \frac{3.33 f_{v,i}^{4/3}}{<r_i>^2} - C f_{v,T}^n}{|\eta_{max}|} \tag{1}$$

The term η_g involves the geometrical effects of precipitates on electron scattering, such as precipitate area and nearest neighbor distance. For η_g the summation is over each precipitate size distribution (for cases involving more than one). The term η_c involves the compositional effects on electron scattering due to fluctuations in the matrix solute content and depends only on the precipitates' total volume fraction ($f_{v,T}$). The variables C and n are positive, non-zero constants. The figure of merit has been normalized by the maximum value of the numerator, η_{max}, to yield a unitless value.

DISCUSSION

Microstructural progression

SEM images of the evolving γ' microstructure with aging are shown in Figure 1. The γ' phase is initially small and closely-spaced, and the precipitates increase in size and interparticle spacing with aging time. A spherical precipitate morphology is seen throughout all aging times.

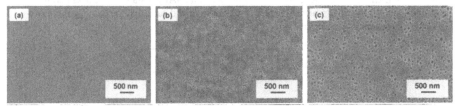

Figure 1. SEM images of Waspaloy aged at 800°C for (a) 8.5h, (b) 38.5h, and (c) 263.5h

Small angle scattering

The effect of the evolving precipitate microstructure on the SAS data is shown in Figure 2. Each subsequent aging time has been shifted in intensity to enhance the readability of the data. The SANS and USAXS data both display visible scattering due to the γ' precipitates that shifts to lower values of Q, and thus larger particle size, with aging time; however, the SANS and

USAXS data do display some differences. The γ' scattering region in the SANS data, displays a correlation peak, which is as expected from interparticle scattering. As the aging time increases, the intensity begins to decrease with Q^{-n} dependence, where n is a positive non-zero integer. This is known as a Debye region. The USAXS data displays Debye regions for all aging times used. The SANS data also displays scattering from γ' at lower aging times than that of the USAXS data, as seen by the large visible scattering region at 0.5h for the SANS data that is absent from the USAXS data. This corresponds to a precipitate diameter of approximately 8nm.

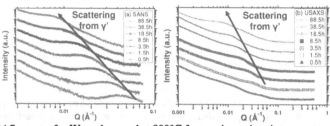

Figure 2. SAS spectra for Waspaloy aged at 800°C for varying aging times as measured by (a) SANS and (b) USAXS. The arrows indicate the shift in the γ' scattering region with aging.

The sample geometries were widely different for the two experiments conducted. For the USAXS measurements, the samples were 10 to 20μm thick, as opposed to 2mm thick in the case of the SANS measurements. The thin USAXS specimens could contain deformed zones from the thinning process that would contribute to errors in the scattering spectra; however, scattering from surface deformation should be negligible for the much thicker SANS specimens. The larger sample volume for the SANS specimens also explains the visible interparticle interference that is absent from the USAXS data, whereby there is a higher probability of secondary scattering as the specimen thickness increases. The earlier detection of the γ' phase via SANS may be explained by a difference in the γ-γ' scattering contrast when using neutrons as opposed to X-rays.

The SAS data were modeled with a volume distribution of scattering objects to yield the particle size distributions (PSD's) shown in Figure 3. The SANS data was not obtained in absolute units, so the PSD's have been arbitrarily normalized such that the data from the two SAS techniques may be compared.

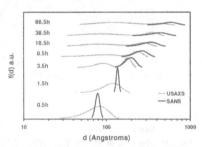

Figure 3. Arbitrarily normalized particle size distributions for Waspaloy samples aged at 800°C for varying aging times as modeled from both SANS and USAXS data.

There are several differences between the PSD's in Figure 3. The initial distributions at low aging times are much tighter in the case of the SANS data than for the USAXS data as a result of better statistics. The more obvious difference between the two data sets is that the SANS data shows no evolution to a bimodal distribution with aging time, as does the USAXS data. The difference in the measured Q range may possibly account for these differences. The maximum measured Q values were ~0.1 Å^{-1}, in the case of the SANS measurements, which effectively may have cut off the scattering from the smallest secondary precipitates that are present at the longer aging times due to the cumulative aging treatments[3].

On the other hand, the primary (large diameter) distributions are shifted to larger diameters for the SANS data. It is plausible that the extensive sample preparation for the USAXS samples may have resulted in the preferential removal of the largest precipitates from the sample surface during the thinning process, effectively skewing the analysis towards the smaller sizes. Therefore, the SANS data, which was obtained from much larger specimens, provided a more accurate measure of the primary precipitate population size.

Figure of merit modeling

The microstructural information required for use in equation 1 was obtained from the PSD's shown in Figure 3. Since the SANS data was not obtained in absolute units, relative volume fractions were used in all cases. The variables C and n were iteratively varied in order to give the best fit of the figure of merit of electron scattering, η', to measured electrical resistivity data from the aged samples. Figure 4 displays the trends in both electrical resistivity and η' as a function of aging time, where η' was calculated for both the USAXS and SANS data.

It can be seen that the empirical trends in the measured resistivity are better captured with the USAXS data, than with the SANS data. Most notably, the peak in resistivity seen at 38.5h is also shown in the η' values obtained by USAXS, but it is absent from those obtained by SANS. As previously mentioned, the resistivity technique is very sensitive to small, critically-sized precipitates on the order of 1nm. The USAXS data was able to display a much larger population of small precipitates at long aging times, because of the wider Q range measured. These results suggest that the smaller precipitate population detected in the PSD's caused the η' values from USAXS data to model the resistivity response more accurately.

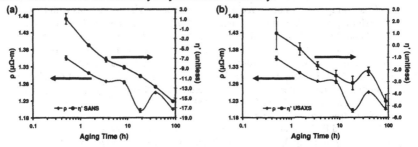

Figure 4. Electrical resistivity and figure of merit of electron scattering as a function of aging time as modeled by (a) SANS and (b) USAXS. It can be seen that the USAXS data results in better empirical trends between the figure of merit and the measured electrical resistivity. This occurred due to a larger measured Q range for the USAXS measurements as compared to that of the SANS measurements.

CONCLUSIONS

USAXS and SANS measurements were used to determine the size distributions of γ' precipitates in Waspaloy samples aged at 800°C for times ranging from a few minutes up to several hundred hours. Analysis of the USAXS spectra showed the presence of a bimodal distribution of precipitate sizes, which is in agreement with the micrographs of samples aged to different times. The secondary precipitates were not detected by the SANS measurements, presumably because of the limited Q range used in the experiments reported here.

A figure of merit of electron scattering, designed to correlate precipitate microstructural changes to the electrical resistivity response, was calculated from both the USAXS and SANS data. The secondary (smaller diameter) distributions detected in the USAXS data were found to provide an increased sensitivity to changes in the measured electrical resistivity, because resistivity is more sensitive to the smallest precipitates. On the other hand, the SANS measurements were able to provide a more accurate measure of the average precipitate size in the primary distribution due to the thicker specimens used.

ACKNOWLEDGEMENTS

The authors wish to acknowledge the funding for this work provided by the U.S. Department of Energy under grant number DE-FG 02-03-ER 46035. Use of the Advanced Photon Source was supported by the U. S. Department of Energy, Office of Science, Office of Basic Energy Sciences, under Contract No. DE-AC02-06CH11357. SANS data was obtained at Oak Ridge National Laboratory's High Flux Isotope Reactor, sponsored by the Scientific User Facilities Division, Office of Basic Energy Sciences, U.S. Department of Energy.

REFERENCES

1. R. J. White, S. B. Fisher, K. M. Miller, and G. A. Swallow, "Resistometric Study of Aging in Nimonic Alloys (1). PE16," *Journal of Nuclear Materials*, vol. 52, pp. 51-58, 1974.
2. V. S. K. G. Kelekanjeri, L. K. Moss, R. A. Gerhardt, and J. Ilavsky, "Quantification of the Coarsening Kinetics of γ' Precipitates in Waspaloy Microstructures with Different Prior Homogenizing Treatments," *Acta Materialia*, vol. 57, pp. 4658-4670, 2009.
3. R. L. Whelchel, V. S. K. G. Kelekanjeri, R. A. Gerhardt, and J. Ilavsky, "Effect of Aging Treatment on the Microstructure and Resistivity of a Nickel Base Superalloy," *Metallurgical and Materials Transactions A*, Submitted for publication.
4. J. Ilavsky, P. R. Jemian, A. J. Allen, F. Zhang, L. E. Levine, and G. G. Long, "Ultra-small-angle X-ray scattering at the Advanced Photon Source," *Journal of Applied Crystallography*, vol. 42, pp. 469-479, 2009.
5. Ken C. Littrell, private communication.
6. J. Ilavsky and P. R. Jemian, "Irena: tool suite for modeling and analysis of small-angle scattering," *Journal of Applied Crystallography*, vol. 42, pp. 347-353, 2009.
7. V. S. K. G. Kelekanjeri and R. A. Gerhardt, "A closed-form solution for the computation of geometric correction factors for four-point resistivity measurements on cylindrical specimens," *Measurement Science and Technology*, vol. 19, pp. 025701-025712, 2008.

AUTHOR INDEX

SUBJECT INDEX

Printed in the United States
By Bookmasters